CLOCK REPAIRING

AND MAKING

A PRACTICAL HANDBOOK
DEALING WITH THE
TOOLS, MATERIALS AND METHODS USED IN CLEAN-
ING AND REPAIRING ALL KINDS OF ENGLISH
AND FOREIGN TIMEPIECES, STRIKING AND
CHIMING CLOCKS, AND THE MAKING
OF ENGLISH CLOCKS

BY

F. J. GARRARD

WATCH MAKER TO THE ADMIRALTY
AUTHOR OF "WATCH REPAIRING, CLEANING, AND ADJUSTING"

WITH ONE HUNDRED AND TWENTY ORIGINAL ILLUSTRATIONS

PREFACE

THIS book is not a history of clocks nor a theoretical treatise, but aims at giving precise practical directions for cleaning, repairing, and making all kinds of clocks. The Author has tried to describe in detail every operation, however small, so that the amateur mechanic with a few tools and no previous knowledge of the subject may, if he so desires, learn from its pages the art of clockmaking.

At the same time, its usefulness is by no means confined to the amateur. Much of the information given and many of the illustrations deal with matters that are not treated upon in any other work known to the Author, notably the descriptions of striking work, ting-tang quarter chimes, tube chime-clocks, and the hints on clockmaking.

All the information given is the result of the Author's own practical experience, and the drawings and photographs are from actual mechanism.

F. J. GARRARD.

Nottington, Weymouth,
1913.

TABLE OF CONTENTS

vii

CLOCK REPAIRING AND MAKING

CHAPTER I.

INTRODUCTION.

DESCRIPTION OF A SIMPLE CLOCK MOVEMENT.

THE illustration on this page (Fig. 1) shows front and side views of a "Skeleton" timepiece. The mechanism of such a clock is simple, and owing to the skeleton form of the frame plates, the wheelwork being easily seen, it is a very suitable

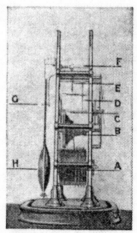

FIG. 1.—Skeleton Clock Movement.

B

form of clock to take as an example in giving a general
description of a spring-driven pendulum timepiece, and an
explanation of the principles of its construction.

The principal part of this clock is the pendulum (H, Fig. 1),
shown more clearly in Fig. 2.   That is to say, the pendulum is
the actual timekeeper.   The rest of the clock merely serves to
keep the pendulum vibrating and to record the number of its
vibrations upon the dial.   If a fairly heavy pendulum is set
swinging, its swings, or "vibrations" as they are termed, occupy
practically equal times, whether they
are long swings or such small ones as to
be scarcely perceptible.   The times are
not *exactly* equal, and in very high-class
clocks this has to be taken into account,
as will be described further on in this
book in connection with compensation
pendulums and regulator clocks.   But
for ordinary house clocks the times
may be taken as equal.

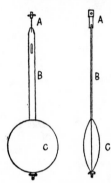

FIG. 2.—Pendulum.

A heavy pendulum, once set vibrating through a fairly large arc, will continue to vibrate for a long time, each
vibration being a little less in extent
than the preceding one, until it comes to
rest.   The causes which bring it finally
to rest are mainly two.   Friction at the point of suspension, and
the resistance of the air to its passage to and fro.   Friction is
reduced to a minimum by employing a suspension-spring,
(A, Fig. 2).   This is a thin, flexible band of steel, or steel
"ribbon."   Resistance of the air is decreased by making the
rod and "bob" (B, Fig. 2) of heavy material, so as to get a
heavy pendulum of small dimensions.   Brass, steel, and lead
are the usual materials of which pendulums are made.   Sometimes also, as in Fig. 1, the pendulum rod is thin and flat, and
the "bob" lens-shaped, to cut the air easily.

A pendulum made and suspended thus, only requires a
very slight impulse from outside to keep it vibrating
continuously, and this the mechanism of the clock is designed
to give.   The motive power is a steel mainspring coiled up
inside the barrel (A, Fig. 1).   Around the outside of the barrel
the chain or line is coiled and in its turn pulls the fusee round

(B, Fig. 1); attached to the fusee, which is a cone-shaped pulley with a spiral groove around it, is the " main wheel" of the clock, the first driving wheel. This main wheel revolves twice per day and drives the pinion of the centre wheel, C. This centre pinion has eight leaves, and as the main wheel has ninety-six teeth, or twelve times as many, the centre pinion revolves twelve times to once of the main wheel, or once per hour. This pinion is therefore prolonged forward through the centre of the dial and carries the long or " minute" hand of the clock. The centre wheel (or second wheel) drives the pinion of the third wheel (D, Fig. 1). The third wheel drives the pinion of the escape wheel (E, Fig. 1). The numbers of the teeth of the

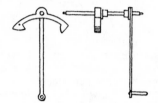

Fig. 3.—Pallets and Crutch.

centre and third wheels and the pinions they drive are such that the escape wheel revolves once in about ⅓ minute. Then having a pendulum of a convenient length for the height of the clock, and making about three vibrations per second, the escape wheel is allowed to revolve just fast enough for the centre pinion to make one revolution per hour.

By means of the pallets and crutch (Fig. 3), each double vibration of the pendulum allows one escape wheel-tooth to pass, and in passing, each tooth gives the pendulum a slight impulse or push forward. Thus the power of the mainspring transmitted through the train wheels keeps the pendulum vibrating, and the pendulum regulates the speed of the escape wheel and so controls the rate of motion of the hands. A short pendulum vibrates more quickly than a long one, and so to regulate the clock, the bob is made to slide upon the rod, and by means of a nut at the lower end (Fig. 2) the bob can be raised or lowered for purposes of regulation. Lengthening the pendulum makes the clock go more slowly, and shortening it makes it go faster.

In the front view of the clock (Fig. 1) several wheels will be seen under the dial—the hour wheel and two smaller minute wheels. These are to drive the hour hand and make it go round the dial once to every twelve revolutions of the minute hand. One minute wheel, sometimes called the cannon pinion, from its shape, is mounted friction-tight upon the prolonged end of the centre pinion of the clock. The other minute wheel, the same size and number of teeth, is driven by the first, and has a pinion of six leaves. This pinion drives the hour wheel of seventy-two teeth and so causes it to make one revolution to twelve of the minute pinion. Fig. 4 shows the motion work. A = minute hand, B = hour hand.

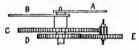

A = Minute hand. B = Hour hand.
C = Hour wheel. D = Cannon pinion. E = Minute wheel.

Fig. 4.—Motion Wheels.

These clocks are generally very well made and last many years. They go eight days and as a rule keep time to within about one or two minutes per week. The regularity with which they run through the week is largely due to the presence of the chain or line and the fusee. It would be easy to omit these parts altogether and place the main wheel upon the barrel direct as in German and French clocks. Such barrels and main wheels combined are called " going barrels." Their disadvantage is that when the spring is wound up at the commencement of the week, the power is excessive, and when it approaches the end of the week, the power is much less. That is to say, the power of such a clock varies throughout the week, causing in the early part of the week excessive wear and tear on the teeth, pinions and pivots, and also tending to cause the clock to gain at first and gradually go more slowly as the end of the week is reached, thus giving an irregular rate from day to day.

The fusee is to equalize the power. When the clock is fully wound up and the mainspring is at its strongest, the chain pulls upon the smallest diameter of the fusee and has little leverage. As the days pass, the chain unwinds from the fusee and gradually pulls upon a larger and larger diameter of the cone until at the last day of the week, it exerts its force upon the largest diameter of the cone. So it will be seen that if the shape of the cone is proportioned to the variation in power of the spring, the pressure upon the working parts of the clock and the force that reaches the pendulum through the pallets, will remain the same throughout the week's run.

It is perfectly true that the time kept by the pendulum is independent of the extent of its vibrations, and so the variations in power should not affect the timekeeping. But this is only true of a *free* pendulum. With a pendulum and pallets such as this clock has, the pendulum is not quite free. It is always being either pushed or retarded by the pressure of the escape wheel teeth upon the pallets. The pressure thus applied tends to quicken the vibrations, and the greater the pressure, the more will they be quickened. So in an eight-day clock especially, the more even the driving power, the better will be the time-keeping and the longer the clock will wear.

A falling weight gives a constant force and is therefore preferred to a spring in high-class clocks. The drawbacks of a weight-driven clock are that the weights require room to fall and necessitate a long case, and also such clocks are not so portable and are heavier.

Another large class of clocks have no pendulums at all, but are governed by a balance and spring after the manner of a watch. The principle is the same. A mainspring and train of wheels finally drives an escape wheel, which through a pair of pallets and a crutch or " lever " gives the impulses necessary to keep the balance vibrating. A balance wheel with a " hairspring" attached acts in a very similar way to a pendulum. If set spinning or vibrating backwards and forwards, it will continue until it is brought to rest like the pendulum by the friction at its pivots and the resistance of the air. Also its vibrations, whether long or short, are performed in equal times. So it fulfils exactly the same part as a pendulum, but is, when well made, much more delicate, and when cheaply made, not so even in its action as a pendulum. Its great advantages are portability and the small space occupied. Its disadvantages are, in good clocks, delicacy and expense ; and in cheap ones, poor timekeeping.

CHAPTER II.

CLEANING A SKELETON CLOCK.

CLOCKS require cleaning for two reasons ; the oil may have dried up or become sticky, or there may be such an accumulation of dirt in the pinions and around the pivots, that it cannot continue to go. Clock oil even of the very best cannot be relied upon to remain fluid more than three or four years, though it may do so in some cases. If, through the clock standing in a hot place, such as a kitchen mantelpiece, the oil actually dries up, and the clock continues to go, as many will if they have good mainsprings and are otherwise in good order, the pivots running in dry brass pivot holes rust rapidly. The friction seems to rust them, aided by the small amount of moisture which there is even in the driest air. Once a pivot begins to rust, its destruction is rapid. The rust powder, oxide of iron, is a sharp cutting and polishing material, and as the pivot slowly and continuously revolves, it is surely ground and polished away to a mere thread. Just how long it will go on before the resistance to motion stops the clock, depends principally upon the power of the mainspring. It is well for a clock that this power is strictly limited, or in many cases, clocks would go until their pivots were all completely worn away before their owners recognized that they required cleaning.

It is best for the clock when the oil gradually hardens and becomes thicker and thicker until it is so sticky that instead of lubricating the pivots and making them turn with less friction, it actually sticks them tight in their pivot holes so that the mainspring is no longer powerful enough to turn them.

So it will be seen that a clock in an air-tight and dust-tight case into which not a particle of dirt can enter, will yet require cleaning merely by the lapse of time causing deterioration of the oil, though its plates and other parts may still exhibit a high polish and its pinions may not hold a spec of dust. It is hard to convince some owners of clocks that this is so, especially in the case of a skeleton clock, or a carriage clock under a glass shade or in a glass case, every part of which can be seen in a highly polished and obviously clean condition.

In either case, whether the oil has dried up, or the clock is dirty, the remedy is cleaning and the process is the same. It must be taken completely apart, the parts cleaned, and it must be put together again and supplied with fresh oil.

Taking the Clock Apart.—The frame of the clock consists of two "skeleton" plates held together by four or six "pillars." The pillars may be either riveted or screwed securely into one plate, and the other plate may be held on by screws

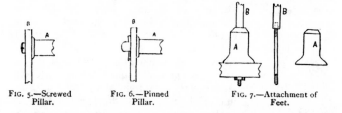

FIG. 5.—Screwed Pillar.   FIG. 6.—Pinned Pillar.   FIG. 7.—Attachment of Feet.

into the pillar ends as in Fig. 5 or pinning through as in Fig. 6. These are the usual methods of holding clock frames together. In Figs. 5 and 6, A = pillar and B = plate. This frame contains the barrel, fusee, and wheels, etc., and upon it are mounted the dial and motion wheels, and from its upper part the pendulum is suspended as shown in Figs. 1 and 2. The complete frame with all its attached parts is known as the "movement" of the clock, and generally stands upon four turned brass "feet," which are held to the base, of wood or marble, by screws and nuts, as in Fig. 7. A is one foot and the figure shows the method of attachment. A long steel screw is firmly fixed into the lower extremities of the clock frame, and passes through the loose foot, and the base or stand, and the whole is held together by the nut underneath.

The pendulum should be first removed by lifting it off the

back cock, or little brass bracket which supports it.   Then take out the pin that holds the minute hand on, and remove the minute hand.   The hour hand is fastened by a little screw. Remove this with a watch screwdriver and take off the hand. Next take off the dial, which is a ring of metal with three or four feet which enter holes in the front frame plate and are secured by pins like Fig. 6.   The dial off, the minute cock can be unscrewed and removed.   This is the little brass bridge screwed to the front plate which holds the minute wheel in position.   The minute wheel and hour wheel (Fig. 4) can then be taken off.   The bridge that holds the hour wheel (if the clock has one) can be unscrewed and the other minute wheel or "cannon pinion" taken off.

The movement, being now stripped of all external parts, may be detached from its base by unscrewing the nuts (Fig. 7) and taking off the loose feet.   Then if the oil has dried up, put a little fresh—any oil will do for this purpose—upon the pivots of all the wheels to make them run smoothly.   Unscrew the back cock and

FIG. 8.—Barrel Ratchet and Click.

remove the crutch and pallets (Fig. 3).   If the clock is wound up, the train wheels will run unimpeded until the power of the mainspring is spent and the clock is run down. In this clock, the train should run until the line or chain is all off the fusee and wound around the barrel.   Although the train can run no further, the line is still tight and the mainspring is wound up a little inside the barrel.   This must be "let down" by loosening the screw of the "barrel click" (A, Fig. 8), then place a large key or a small hand-vice upon the square of the barrel arbor, and easing the click from the ratchet teeth B, allow the square to run back until the spring is quite unwound and the line becomes slack. There being now no power on the clock, the plates can be taken apart and all the wheels, etc., removed.

Cleaning.—Petrol is the best cleansing fluid to use for all clock and watchwork.   Great care has to be taken in its use as it is highly inflammable, and it is best when using it in an open bowl to do so in a room with no artificial light burning, or at least far from such a light, and to only leave it exposed for a short period.   An enamelled iron hand-bowl is useful for clock cleaning, and into such a bowl, the plates and all wheels, cocks, pallets, etc., may be put, and about half a pint of petrol poured upon them.

The fusee can be taken apart by removing the cap on the main wheel, and the main wheel, fusee body, ratchet and click may be put in the petrol with the other parts. That is, if the clock has a chain. If it has a gut line, do not put the fusee body and line into the petrol, but be content to wipe it clean with a paper and a duster.

The barrel cover can be prised off by means of a screwdriver and the arbor removed, cover and arbor being put in the petrol, but not the barrel and mainspring. It is not often necessary to take the mainspring out of its barrel, only when it is broken, or when the oil upon it has become so sticky that the coils adhere to one another and prevent it acting properly. If the spring has to be taken out, put it and the barrel in the petrol as well, but if not, do not let the petrol touch it, as it is impossible to remove the petrol from the coils while the spring lies in the barrel, and its presence has a bad effect upon any fresh oil that is applied.

The parts in the hand-bowl may be taken one by one and brushed clean from dirt and old oil by means of a stiff clock brush and petrol. As each part is cleansed, shake off the surplus petrol into the bowl and dry it with a duster. Serve all the parts in the same way and the petrol will be observed to have become quite dark and dirty with the dissolved grease and dirt from the clock. Still it can be used again and again, until quite saturated with grease, so it may be allowed to settle for a moment to let grosser dirt sink to the bottom, and then quietly poured into another bottle for the next clock. A tin funnel is useful for the pouring. Clean petrol, for a particular job is thus kept separate.

If it is deemed necessary to take the mainspring out of its barrel, hold the barrel in a duster to protect the hand, and seize the centre coil of the spring with a strong pair of pliers and pull it up and out. This is assisted by, at the same time, turning the centre coil a little in the direction of winding up, which tends to ease it. A few pulls in this way at the centre coils will get the spring out. Care has to be exercised to prevent it springing out all at once and injuring the fingers, hence the use of a duster to grasp it.

When out, the spring can be wiped clean from end to end after being washed in petrol.

This simple process of washing in a bowl cleanses the clock only. If it is desired to re-polish the plates and wheels as well, as is the case if they are tarnished or corroded, they must be submitted to another process before putting in the petrol bowl. Tarnish may be removed and a high polish given by using " Globe " metal polish on a fine brush or a rag, a thorough washing in petrol following to remove the dirt. Or rottenstone and oil on a brush may be used in the same way, but it does not give such a high polish, and in the case of a clock every part of which is visible under a glass shade, the " Globe " polish is to be perferred. Rottenstone is useful when the plates are corroded and require a sharper cutting medium. When badly corroded, bath-brick dust may be used first; in both cases finishing with the " Globe " polish.

The parts having been polished and cleansed by which-ever process is deemed most suitable, may now be taken in hand for more particular treatment. Each pivot hole must be " pegged " out clean. Bundles of " pegwood " for this purpose are bought at clock-tool shops and are little sticks of dog-wood about five or six inches long. Take one of these and sharpen it with a pocket knife to a long thin point. Insert it into a pivot hole and turn it round a few times. It will come out black and dirty with the old oil and the polishing medium used to polish the plates. Re-sharpen the peg and repeat the process until each pivot hole is absolutely clean. The larger holes such as the fusee and barrel holes may be cleaned out with first a duster and then a leather wrapped around a peg. After pegging out thoroughly, take each part and hold it in tissue paper while it is polished up quite clean and bright with a stiff watch brush charged with just a little dry chalk dust. Rub-bing the brush once across a white billiard chalk suffices to charge it with sufficient chalk. This should leave the parts beautifully clean and polished, and from this point onwards none of them should be handled with the unprotected fingers, or stains and greasy marks will result. Always use tissue paper to hold them, or pick them up with tweezers. Wheels may be held by the edges of their teeth. If the plates are handled by the *edges only*, the edges may be given a final rub with a leather when the clock is put together the very last thing before putting on the shade.

Putting together.—If the mainspring has been taken out of the barrel, it must first be put in, and a firm grip and a strong wrist are required. Lay the barrel flat upon a wood table and hold firmly in one hand, protecting the hand with a duster. Take the outer end of the spring in the right hand and insert it in the barrel, hooking it securely on the hook. Then coil it in an inch or two at a time, pressing each coil well down, and never once relax the grip until all is in, or it will fly out. As each coil is pressed in, turn the barrel in the left hand to bring the next part into position for the right hand to press it in. When all is in, bang the barrel hard upon the table two or three times to shake the spring well down to the bottom. Insert the arbor to see that it hooks properly. Apply good machine oil to the coils of the spring and the barrel bottom and put on the cover again. Oil each pivot of the arbor and then polish up the barrel outer surface to look nice and remove all finger marks.

The fusee body, ratchet, click and main wheel all being clean, put on the ratchet, put in the click, and oil the click with clock oil (purchased in small bottles at the clock tool shop). Oil the fusee arbor where the main wheel turns upon it, and the groove where the cap or key fits. Then put on the main wheel and oil where it rubs against the fusee body, put on the cap and insert the pin or screw that holds it.

Now the plate with the pillars fixed in can be laid upon the board, and the barrel, fusee, and wheels all stood up upon it in their pivot holes. Put on the top plate, and pressing it down with tissue paper, insert one pillar pin at the bottom. Then pressure can be exerted upon the other end of the plate as each pivot is got into its pivot hole until all are in and the top plate goes on flat. Put the pillar pins in quite tight and press home well with pliers. The chain or line can next be put on. First hook it into the barrel, and guiding it with the fingers, wind it up regularly, following as far as possible the old marks. When all is on, hook it in the fusee and continuing to turn the barrel, draw the chain tight. Put on the barrel ratchet and click (Fig. 8), and grasping the square arbor, wind the spring up one tooth of the ratchet just to keep the chain tight. Then rearrange the coils around the barrel quite regularly and finally wind the spring up half a turn more and screw the click tight. This is termed "setting up" the mainspring.

Put in the pallets so that the train cannot run, and proceed to wind up the clock, closely watching the chain to see that it goes straight on to the fusee and does not drag sideways or it will get out of its groove and damage the edges, or if it is a gut line, will cut the line. If any sign of sideway dragging is observed, bring pressure to bear upon it as it is wound by something smooth, such as a burnisher or a pocket-knife handle, to keep it quite straight. Then apply a *little* clock oil to all pivot holes and to the pallet faces, and the clock if in good order should "trip" merrily without the pendulum.

The frame or movement can be once more fixed upon its stand and the feet put on. There will only remain the dial, hands, and motion work. Clean the minute wheels, hour wheel, and minute cock in the same way as the plates, etc., and put them in position. In some of these clocks the motion work is arranged like that of an eight-day English dial, and in others it is somewhat after the fashion of a French clock.

In the English form (Fig. 9) the minute wheel, known as the cannon pinion, fits loosely upon the centre arbor and rests

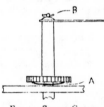

upon a short curved spring A. When the minute hand is put in place and a washer or "collet" put on over its centre, the whole is pressed down against the bow-spring A and a pin B inserted to keep it down. The cannon pinion is then held friction-tight between the bow-spring A and the pin and washer B, so that it is carried round upon the centre arbor as the clock goes,

Fig. 9.—Sprung Cannon Pinion.

and when it is required to set the hands it may be turned stiffly ; care should always be taken to see that the spring A does not touch the plate. This is a very reliable method of carrying the minute hand, as the tightness may be regulated to a nicety by thinning the spring to ease it, or hammering it stiff and bowing it a little more to tighten it. The hands should just move stiffly enough not to drag behind as the clock goes, but must not move so stiffly as to cause danger of breaking the minute hand when setting them. In this type of motion work the hour wheel is generally carried by a fixed brass bridge screwed to the front plate and spanning the cannon pinion without touching it. Movements arranged thus are known as "bridge movements," and are considered better and more reliable than the pattern without a bridge.

Fig. 10 shows this latter form. In it the cannon pinion fits rather tightly upon the centre arbor and has a spring A to give it an even frictional motion. If the cannon pinion merely were tight upon the arbor, it would turn in a jerky manner and after a little while get so easy that the clock would go while the hands remained stationary. To prevent this, a portion of the "pipe" at A is filed thin and pressed inwards a little to give the pipe a spring grip upon the arbor. In this pattern, there being no bridge, the hour wheel rests and turns upon the pipe of the cannon pinion. It is a neater and simpler method of arranging the hand work, but rather more liable to give trouble by working loose than a "bridge movement" would be.

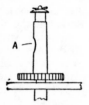

FIG. 10.—Snapped Cannon Pinion.

The minute wheel pivots require a little clock oil, but none should be put upon the pipe of the hour wheel. When the motion work is on, put on the dial and pin its feet securely. The dial will probably be french silvered on brass and the figures painted. If so, the only cleaning permissible will be a gentle wipe with a soft leather or duster. The hands may be put on, and the pendulum polished up and re-hung. A little oil should be put upon the pin of the crutch, and the slot in the pendulum rod where it works must be scrupulously cleaned and burnished bright.

The clock will now be complete and only needs setting in place again and care taken that it is perfectly "in beat." As this is a process always required in a pendulum clock, it will be here described. A pendulum clock is in beat when as the pendulum hangs motionless and at rest, an equal movement in either direction causes an escape tooth to drop. In other words, a "tick" should come equal distances on either side of the position of rest. The readiest way to tell if a clock is in beat is to start it with the least movement of the pendulum that will allow it to go, and listen carefully to the "tick" or beat. If quite even thus :—tick, tick, tick, tick, it is nearly enough " in beat." But if like this :—tick tick, tick tick, etc., etc., in twos, it is not in beat. By watching the pendulum bob and seeing on which side it swings furthest after the "tick," the crutch wire may be very slightly bent to one side until the beat is right. A little observation and consideration will show to which side the crutch requires bending, or a single trial bend will show at once.

Clock cleaning requires a few tools, such as screwdrivers of suitable sizes, pliers, a small hand-vice, brushes, pegs, etc., and cannot be properly done without them. Screwdrivers that are too large will spoil the screws and mark the brass work, and the absence of proper pliers, etc., will risk disaster in allowing parts to slip at a dangerous moment. Also only good clock oil should be used for pivots and light working parts. Other oil will dry up, get sticky quickly, or corrode the plates green and necessitate recleaning in a few months or perhaps injure the clock as well.

CHAPTER III.

REPAIRING A SKELETON CLOCK.

Broken Mainsprings.—The mainsprings of these clocks do not often break, and when they do it is generally just at the outer end across the hook hole. When this has happened, take the spring out, and trim off the broken end to a rounded form as in Fig. 11, the end being straight with corners well rounded. Then soften the steel for about one inch back from the end, by heating in a gas flame or by a spirit lamp. The spring is hardened steel and heat will soften it. The degree of heat is indicated roughly by the colour of the steel, buff the surfaces bright and hold in the flame. As the steel gets hot it will assume a straw colour, then a blue, this will pass off to a dirty white or grey. At this point stop, as it is hot enough for the purpose and heating it actually to redness will make it too soft. When cool, punch a hole in it by placing the spring on a steel stake with a hole in it and using a flat-faced punch, or by placing on a block of lead or hard wood and using a round or pointed punch. After punching, the hole may be filed out larger and to the exact size and shape required by a rat-tail file. The spring is then put in its barrel again as described in Chap. II.

Fig. 11.—Main-spring End.

A spring that breaks just at its extreme inner end is very troublesome to deal with, and most clockmakers discard them and fit new ones. It is generally impossible to make a really good job of such a breakage. But there are circumstances in which a new spring cannot be obtained just when required, and in such a case the attempt may be made. There are two difficulties, the first is to soften the end without softening the adjacent coils, and the second is to get at it to punch the hole and file up the end. A blow-pipe well directed is sometimes successful in softening the end, but the greatest care is needed to prevent the flame from wandering. Another method is to heat a bar of half-inch round iron rod to redness and insert it into the centre of the spring and apply it to the centre coil as far as necessary. One difficulty is to see that the softening does not end suddenly, or a sharp bend or a break results. The softened part must gradually merge into the hard portion by a gradation of colour extending over an inch or so. Then as to the hole, the centre coil must be forcibly pulled up and held in a vice while a hole is *drilled* in it. The drilled hole can be enlarged by broaching and then filed out with a rat-tail file. When the spring-end is finally completed, the centre coil must be forced down flat again and curled inwards to catch well on the hook upon the barrel arbor. Many springs treated thus last for years again, others are a failure.

A spring broken somewhere in the coils, about the centre

FIG. 12.—Repairing a Mainspring.

of its length, or at least not very near to the "eye" (the middle) can be mended by softening the two broken ends and riveting them together with three or four rivets as in Fig. 12, and such springs last well. Of course their action is not so smooth and regular as a perfect spring, but it is a useful repair on occasions.

It must be understood that when springs break in the eye, or in the coils, in fact anywhere except at the extreme outer end, the *best* repair is a new mainspring, and where one is readily procurable, repairing the old spring should not be thought of.

Broken Chain or Line.—A chain is far better than a gut line, but more expensive, so the best clocks generally have chains and the cheaper ones lines. Chains occasionally break, and to mend them, the old rivets must be punched out, leaving two ends as in Fig. 13, and a new steel rivet filed up, inserted, and cleaned off smooth and flat on each side. The principal difficulty is to get the old rivets out. Just file their outer surfaces a little and placing the chain end over a stake with a hole in it, and using a small flat-ended punch, drive out the rivet by a sharp hammer blow. A new chain hook is simpler, as only one link requires opening and one rivet removing. In Fig. 14 A is a barrel-end hook, and B a fusee-end hook. These hooks can be bought at the clock material shop, but to make

FIG. 13.—Repairing a Chain. FIG. 14.—Chain Hooks.

one is quite an easy matter. A piece of thin flat steel, such as a piece of old broken mainspring, is taken and softened, the rivet hole drilled and then it is filed up to shape and smoothed.

A broken line cannot be mended, as a knot would not pass the fusee groove, and no method of splicing is possible. The lines are sold ready to put on. The method of attaching is the only difficulty. To attach to the fusee, take the fusee to pieces as described in Chap. II. and remove the old broken end. Pass the new line through the hole and draw through inside for a couple of inches. Tie a simple knot as at A, Fig. 15. Gut is peculiar stuff to tie tightly. It tends to loosen itself, so cut off the end with cutting nippers fairly close to the knot, light a match, and hold the edge of the flame just touch-

FIG. 15.—Securing Knots in Gut. FIG. 16.—Securing Barrel End of Gut.

ing the cut end of the gut and sear it. This forms a hard lump on the end and prevents it drawing through as at B. Be careful that the flame does not scorch the knot, but only the stumped end. Then pull the line through and tuck the knot in its hollow, put on the ratchet, etc., and fasten up the fusee again. Fig. 16 shows how the barrel end of the line is fixed. First wind the new line up on the fusee with the fingers until it is filled, and allow four inches to spare. Cut it off here, and insert the end into the three holes in the barrel as shown, in hole No. 1 out of hole No. 2 in hole No. 3, and under and through the inner loop A. Draw the end through, cut off and sear with a match as before. Then push the gut back as far

C

as it will go, and work the knot tight, taking up all slack. Finally well oil the line from end to end, it keeps out damp and preserves the gut.

It may here be remarked that gut lines are best kept in oil, it preserves them and keeps them pliable.

Hands.—Broken hands must be replaced, but *may* be repaired by soldering a little slip of mainspring steel behind them and washing the soldering acid off well with water to prevent rusting, see Chap. IV. If brazing is used a stronger job is made and the hand may be repolished up and blued after the repair. New hands usually require a little filing out of their centre holes to fit them and will need no special instructions.

Rusty hands must have the rust filed right out, resting the hands on a block of cork screwed in the vice. Smooth them with emery buffs, first a No. 1, finishing with about a $\frac{3}{0}$, the finest, this will leave nearly a polish. Be careful not to touch the surface with the fingers if a nice blue is required, and blue them by holding gingerly over a spirit lamp flame and colouring evenly from end to end. This is a tricky job, and if a failure at first, buff off the colour and begin again. A few tries will enable it to be done fairly well. After blueing, to prevent rusting again, if in a damp place, a coat of thin varnish or lacquer may be given.

Broken Wheel Teeth.—Broken teeth seldom occur in these clocks, but are fairly easy to replace. Fig. 17 shows the

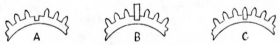

FIG. 17.—Replacing a Broken Tooth.

process in three steps. First file a slot as at A, the width of the tooth and as deep as the wheel rim allows without undue weakening. Then file up a piece of brass and soft-solder in as at B. Wash off well in water to remove the acid, and file up the new tooth as at C.

Rusty or Worn Pinion Leaves.—Rusty pinion leaves

must be scraped by a hard steel scraper such as a graver, or screw-driver blade, and may be scoured out by emery and oil mixed to a paste and applied on a peg as a rubber. A good washing in petrol is required after, to remove all traces of emery. Emery is a good cutting agent, but bad stuff to leave on any part of a clock, it continues to cut and will wear out the parts rapidly if allowed to remain. Worn pinions are generally cut into little hollows just where the wheel works, and as a pinion generally has plenty of length, the wheel can be moved a trifle so as to work in a different spot. This can readily be done by bending each arm a trifle in the direction desired, making the wheel slightly cap-shaped. Another method is to turn the pivot shoulder of one pivot back a little so as to let that pivot further through its hole, and put a collet or washer on the other pivot and so move the wheel and pinion bodily. Or instead of a collet on the pivot, a " raised bush " may be put in the pivot hole (see Bushing Pivot Holes). After treating thus, the wheel teeth will work in an entirely new part of the pinion and the latter will be given a new life. To turn a pivot shoulder back requires either a small lathe or a pair of clockmaker's " turns," information on the use of which is given in Chap. IV.

Worn Pivots.—When a clock runs after the oil has dried up, the pivots wear badly and are often cut to half their original thickness. The cutting or wear is generally uneven, being in ridges, with a kind of mushroom end left on as in Fig. 18. Such pivots must be made smooth and parallel once more. The only proper method is to turn and polish them in lathe or turns as in Chap. IV. But rather than leave them alone, if no lathe or turns are available, lay the pivot in a groove in a piece of boxwood screwed in the vice, and holding the pinion in the fingers, very carefully file them smooth with a fine and sharp file, afterwards smoothing them with a $\frac{3}{0}$ emery stick and finishing with a hard steel burnisher well oiled. They must be kept as round and parallel as possible. After such treatment the pivot holes will require re-bushing to fit the reduced pivots. In renovating worn pivots, they must be smoothed until the marks of wear are gone, but should not be reduced in diameter more than is absolutely necessary, as a

FIG. 18.—Worn Pivot.

pinion has only two pivots, and when they are gone, to replace them is difficult. Still, this point will be reached at last, when a pivot is so rusted or worn that it is too thin for use, or when a pivot is accidentally broken off. Then there is nothing for it but to put a new pinion entire, or drill up the arbor and fit a new pivot. If the other pivot is also badly worn and the pinion leaves as well, a new pinion is the best remedy. But if the pinion leaves and the other pivot are still serviceable, the best course is to drill a central hole up the arbor in a small lathe, or by hand, drive in a plug made of a hardened and tempered steel needle (see Chap. IV.) and upon it turn a new pivot. Fig. 19 shows the process. A shows hole drilled up centre, B the plug inserted and driven tight, and C the new pivot. Before drilling it is advisable to soften the end of the arbor by heating it nearly to redness, so as to be able to drill it easily. When finished the discoloration caused by the heat may be removed by a $\frac{3}{0}$ emery buff stick.

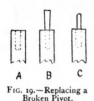

A   B   C

FIG. 19.—Replacing a
Broken Pivot.

Bushing Pivot Holes.—When pivot holes in brass plates wear large or when pivots get worn and have to be re-smoothed and polished, the pivots no longer fit, but have side play that renders the " depths " uncertain. A " depth " in a clock is the action between the teeth of a wheel and the leaves of a pinion which it drives. The teeth of a wheel up to a certain point —" the pitch circle "—are straight radial lines drawn from the wheel centre. Similarly the pinion leaves are straight lines up to the pitch circle. Beyond the pitch circles, the wheel teeth are curved to a point and the pinion leaves are rounded. When a depth is correct, the pitch circles of wheel and pinion roll upon each other. When a wheel is driving, it is the curved portions of the wheel teeth that act upon the straight portions of the pinion leaves. Fig. 20 shows a depth. AA and BB are the imaginary " pitch circles." A depth will only run properly and transmit the power evenly and without loss when pitched like Fig. 20. So it is important that pivot holes should be in the exact position required and that the pivots should fit the holes perfectly.

For bushing holes, " bushing wire " is sold in lengths. It

is round brass wire drawn with a central hole in it like Fig. 21, and can be obtained in all sizes to fit all pivots. Select a piece the hole in which is a little too small to go on the pivot, and with a file slightly taper its end. Then broach out the pivot hole large from the inside of the plate until the bushing wire

FIG. 20.—Wheel and Pinion Depth. FIG. 21.—Bushing Wire.

can be inserted. Then drive in the wire tight, cut it off level on both sides, slightly rivet it, and finish it off flat and level with the surface of the plates by filing, and finally polish it smooth with polishing paste. A new oil-sink can be cut by a large drill blade and finished with a round chamfering tool (see Chap. IV.). A much neater and better way to bush a hole is to use a clock or watch lathe. Hold a piece of bushing wire in a split chuck and turn it tapered and cut off a small length. If the centre hole were always quite true, the bush could be finished at one chucking, but it is not to be depended upon. Therefore, take the small bush thus made in the rough and put it on a turning arbor in the lathe or turns, and turn it true to the centre hole, tapered and hollow on each end face, so as to facilitate riveting the edge over. Then if the hole is opened out just to the right size and the edge bevelled off a trifle, the turned bush can be inserted and riveted in flush and the job will be almost finished, as it will be so neat and smooth, as well as quite true. This method is by far the best as it does not alter the depths which the hand-filing method is apt to do, and it is quite as quick, time being saved on the finishing.

If before bushing a hole, the depths are carefully examined, if one is shallow or deep, it can be corrected at the same time by filing the hole towards the shallow depth or away from the deep depth before finally broaching it out to fit the bush. Then, when bushed, the faulty depth will be corrected at the same time. Always after bushing a hole, spin the wheel by itself between the plates to see if it is quite free and has proper endshape. Each wheel in the clock *must* have just a little endshape or it is liable to tighten up. In finally opening out a bushed hole to fit its pivot, care must be taken that it is upright by sighting the broach from two positions.

Worn Pallets.—The question is sometimes asked, how do brass wheel teeth wear hard steel pallets ?   The answer is that it is not the brass teeth that do it, but the dirt or rust powder. The oil on the pallets collects the minute particles of dust in the air.   This dust is of all kinds, some soft and incapable of doing harm to the steel, other particles are hard and capable of cutting like emery powder.   The mixture finding itself between two surfaces, the brass tooth point and the hard steel pallet, imbeds itself by the pressure of the softer of the two, viz. the teeth points, which then act as polishing tools charged with a cutting medium and wear away the steel pallets. Or when the oil has dried up, and rusting commences, the red rust powder imbeds itself in the same way in the brass and cuts the pallets.   From this it follows that a clock kept free from dust and never allowed to run until the oil dries up, will not show wear upon the pinions, pivots, or pallets ; and such is in fact the case.   Many such clocks after going fifty years show only a brightness upon the steel surfaces.   Clocks are worn out by dirt and rust, and not by the wear and tear of going.

When pallets are worn slightly, the wear marks can be buffed out with emery sticks, finishing with a $\frac{3}{0}$, using strokes in the direction of wear of the pallets and not across them.   An emery buff, however fine, leaves a slight " grain " on the steel like parallel lines or ridges in the direction in which the buff is used, and if these ridges go across the pallets, they leave a fine ribbed surface that is apt to act as a fine file, and not only increase the friction, but wear away the teeth points.   So always buff in the direction of wear of the teeth points.

If the pallets are deeply worn, buffing out the wear marks will so alter their shape that the depth will be too shallow and the back pallet pivot hole will need drawing nearer to the escape wheel and bushing as described earlier in this chapter. In such a case it will be best to move either the escape wheel or the pallets along their arbors so as to bring the action of the teeth on to a fresh portion.   The pallets are generally wide enough to allow of two such moves, one on either side of the first or central position.   Some pallets are merely driven on their steel arbors and can easily be moved.   Others require the brass seating moving bodily. The brass collet on which the pallets are seated is generally soft-soldered on the arbor and can be warmed and moved a little. Or the arbor may be moved bodily by a raised bush, or the escape wheel may be bodily moved by the same means, and if its pinion is worn as well, which not infrequently happens, the one move will correct both faults.

A pair of pallets will finally reach such a stage that after being shifted for wear several times and after being repeatedly buffed smooth, they are still deeply worn and the depth is very shallow. Then resort can be had to "facing." This is really a very good way of treating them, and consists of soft-soldering two pieces of watch mainspring on their wearing faces. The pallet surfaces are first buffed bright and clean, and "tinned" with solder over a spirit-lamp flame. The two little pieces of watch mainspring are buffed bright and "tinned" also. Then one by one they are laid on the pallet faces, warmed until the "tinning" melts, and pressed down close till set. A good washing in water will prevent rusting. The surplus mainspring is trimmed off all round by filing and buffing, and the acting surfaces of the faces polished with a $\frac{3}{0}$ buff. This makes a thoroughly good job. The surfaces are then blue-tempered steel instead of dead-hard steel, and will wear a little more quickly, but, on the other hand, they can be restored again and again by soldering on fresh pieces of spring, which is only the work of a few minutes. Fig. 22 shows a pallet thus "faced."

FIG. 22.—Facing Pallets with Watch Mainspring.

The escapement (wheel and pallet action) of a clock is a most important part and should be carefully studied. The form of escapement used in these clocks is the "recoil," so called because when an escape wheel tooth has "dropped" upon a pallet face, the continued motion of the pallet in the same direction causes a recoil or backwood motion of the escape wheel. Fig. 23 shows the escapement. The escape wheel travels in the direction of the arrow, and the pallet a tooth first meets, A, is called the "entering pallet,"

FIG. 23.—English Recoil Escapement.

the other, B, is the " exit pallet." The pallet faces are curved and the backs may be straight or curved, it is immaterial so long as they clear the backs of the teeth when in action. To be about right (and much latitude is allowable), a line drawn along the face of pallet A, as shown by the dotted line, should be horizontal, and a similar line across the face of pallet B, vertical. Then the impulses and recoils will be at about the right angles. The "impulse" is the amount by which the pallet intersects the circle of the wheel teeth, and is really the distance through which the teeth push the pallets and through them, the pendulum. When pallets wear and the depth becomes "shallow," the intersection is small and the impulse diminished. If this is diminished beyond a certain point, the pendulum will not receive sufficient impulse to keep it vibrating.

When a tooth point travels along the face of pallet A, it forces the pallet upwards and gives impulse to the pendulum. When the tooth reaches the point of the pallet it " drops " off and a tooth further round the wheel "drops" on to the face of pallet B. As the pendulum swings back, this tooth gives impulse to pallet B by pushing it outwards out of its path until it reaches the point of the pallet and "drops" off, another tooth then falling or "dropping" on to pallet A, and so on. So the distance moved by the escape wheel at each swing of the pendulum is half a tooth-space and is made up of a long impulse during which it drives the pallet before it, and a short "drop" during which the wheel simply travels forward free until the next tooth comes into contact with the pallet face and is arrested in its motion. The "impulse" is work done, i.e. power transmitted to the pendulum, but the "drop" is power wasted, i.e. it does nothing towards keeping the pendulum going. It follows that the smaller the "drop" and the longer the "impulse" the better the clock will go. A little "drop" is necessary to ensure that the pallet points do not catch the backs of the escape wheel teeth. More than this is wasted power. So in these and all escapements, the depth should be adjusted so that the "drops" are as small as it is safe for them to be and they should be equal on each pallet.

If there is too much " drop" on to the face of A, it shows that the tooth escapes from pallet B too quickly, or that pallet

B is too short. This can be remedied by bringing the pallets bodily lower down and nearer to the wheel. If a tooth "drops" too much on pallet B it shows that a tooth escapes from the point of pallet A too soon. To equalize matters, buff a little more off the point of pallet B, which has the effect of giving more "drop" upon A also. Then *both* pallets will have too much "drop" and it can be remedied by bringing the pallets a little nearer to the wheel as before.

Damaged Escape Wheel Teeth.—The teeth of an escape wheel should be thin at the points, all of exactly equal length, and equally spaced. Wheels of the pattern used in these clocks have teeth with curved front surfaces and straight backs. When a new wheel is cut, the backs are all equally spaced and wear of the points only affects the length and not the spacing of the teeth. So in repointing up injured teeth, file only on the curved surfaces and leave the backs alone. This will leave the wheel equally spaced, but it may have some short teeth. Such a wheel needs "topping." To "top" a wheel, run it in the turns or between centres in a lathe, and taking a very fine file, and steadying it on the T-rest, lightly touch the teeth points as the wheel revolves rapidly under it. This cuts them all down level and will leave little flats upon those that were longest. These will require a little careful filing with a curved file ($\frac{1}{2}$ round, or "crossing file") on their curved surfaces to point them up. After "topping" a wheel the depth will be made a little shallow and the pallets will require bringing nearer to the wheel again by drawing and bushing the pivot hole or holes. Another way of deepening a pallet depth is to lower the back cock which holds the back pallet pivot. This can be done by bending its steady pins upwards and drawing the screw holes a little. Or for a gross alteration, the steady pins may be filed off, the screw holes filed oval, and the depth adjusted until correct, then the cock may be re-steady pinned in a fresh place.

Pendulum Suspension Spring, etc.—A new pendulum suspension spring may be either bought, or made from the thinnest watchspring. A stiff watchspring is too stubborn for these clocks. The spring should be straight and true. A "kink" in it is fatal and will cause the pendulum to "wobble" as it swings and stop the clock. The spring should be quite tight in the top of the pendulum rod. In case of breakage and loss, the length of the spring may be ascertained by noting where the pin of the crutch worked in the pendulum slot. It will have made a mark. Then make the new spring to bring the pin to the old mark again, otherwise it may not be possible to regulate the clock.

If the rating nut at the lower end of the pendulum does not "bite," a new nut may be made a little smaller in the hole and the end of the rod re-tapped. Hammering up the nut to tighten it is of little use. Or a new tapped end may be fitted to the pendulum. It is merely a piece of brass soft-soldered into the pendulum rod and is soon replaced.

The crutch pin must be smooth and burnished and quite firm in the crutch. The crutch also must be firm upon the pallet arbor, or much of the impulse will be lost. The pin should be quite free in the pendulum slot, but should have little or no side play in it. The slot itself should be smooth and well burnished inside, and slightly rounded to prevent binding if the pendulum rolls a little.

CHAPTER IV.

SPECIAL TOOLS AND PROCESSES.

In the foregoing chapters references have been made to various tools, etc., used by clockmakers, and to several processes, such as soldering, hardening and tempering steel, etc. Before proceeding further it will be best to describe them.

Hardening and Tempering Steel.—Steel possesses the peculiar property of being capable of many degrees of hardness, and it is by taking advantage of this fact that tools are made to cut. Thus soft steel can be cut by tempered steel, and tempered steel by dead-hard steel. Steel rod bought from the metal warehouse is soft or in its natural state. It can be made a little softer by heating to redness and allowing to cool slowly, or by heating to redness and allowing to cool until no red is visible in the dark and then plunging into water. This is termed "annealing." In this soft state it can be easily bent, cut, filed or turned, and tools are made by this means. When a tool such as a lathe cutter or drill is thus made from soft steel, before being of any use, it must be hardened, and this is done by heating to redness and plunging into water. High grade steel such as tool steel will harden at a dull red heat, medium steel must be made a bright red, and mild steel requires to be made white hot. But the less the steel is heated, the less it is burned on the surface and the tougher it will be when hardened. So tools are made from steel of a high grade.

When thus hardened, the surface black scales off into the water in patches and shows a white surface underneath. This is a sign that the steel has been successfully hardened. A conclusive test is to try to file it. If soft, it can be filed; if hard it cannot, as the file itself is only hard steel, and both being equally hard, the steel cannot be cut. But in this "dead-hard" or "glass-hard" condition it is far too brittle for most purposes. It is only useful for cutting or drilling tempered steel. For any other purpose it must be "tempered." Tempering consists of heating it again gradually and is really a partial softening process. A dull red heat, the hardening heat of good tool steel, is about 1000° Fahr. If after hardening as above described, the steel is tempered by heating to half this heat and allowing to cool, a sort of semi-softening is obtained which renders the steel about mid-way in hardness between "soft steel" and "dead-hard steel." As a clean bright piece of steel is heated gradually, the surface shows certain colours in succession, and the heat and degree of "temper" can be judged by watching the colours. It first assumes a pale straw tint, then darker, and purple, dark blue and light blue in succession. A pale straw tint indicates a temperature of 450° and is the temper for punches, drills, lathe cutters, taps, etc. At this heat the hardness does not suffer much, but the brittleness is decreased and some elasticity given. Purple or dark blue indicates about 560° and is the temper for springs, screws, etc., and steel parts in general that are required to be tough, elastic, and to wear well, and includes all pinions, pivots, and arbors in a clock. A dark blue is the temper that gives steel its maximum of elasticity. But this temper is too soft for any cutting tools. Straw-tempered steel can hardly be touched by a file, and a few strokes would completely spoil the file. Blue-tempered steel can be turned nicely by hard, sharp tools and filed with care if fine-cut files are used, but it is difficult to drill and impossible to tap, *i.e.* cut screw threads in or upon. It would simply spoil the screw plates and taps.

When tempering it is advisable to have a dish of water handy in which to plunge the steel as soon as the colour indicates the correct heat, as the heat travels along the metal and is apt to make it softer than is desirable. It is immaterial as far as results are concerned whether a piece of tempered steel is allowed to cool slowly or is cooled by dipping in water. The water is only useful to prevent the heat increasing further.

An example or two will be given. Suppose a drill has been made from tool steel rod as described later in this chapter, it is held in a gas flame until the blade end is red-hot.

Then it is withdrawn *instantly* and quenched in water. It is tested for hardness, and if hard, tempering is taken in hand. In hardening steel there are no intermediate degrees of hardness. Either the steel is hard or it is quite soft. If red-hot when dipped in the water it will be hard, hence the quickness necessary when hardening a small article like a drill. Passing through the air from the gas flame to the water cools it, and it is not desirable to over-heat in the flame, so the water should be held as close as possible to the gas flame, and the drill withdrawn from the flame and plunged in the water with a very quick motion. So in testing for hardness, no doubt need be entertained. If it can be filed, it is not hard. But an inspection will show with equal conclusiveness. If scaled white it is hard; if not, it is still soft and must be tried again at a slightly greater heat.

When satisfied that it is hard, take an emery buff and brighten one flat so that the colour may be watched. Then have a cup of water handy close to the flame ready to dip, and hold the drill stem just behind the blade, in a spirit-lamp flame, and watch the colour. It will become a straw tint in a moment or two and the tint will travel towards the drill point. Instantly plunge it in water to arrest the heat. The result is a straw tint at the drill point or blade, getting darker as it goes down the drill stem, and so giving a hard edge and an elastic stem that will not readily break or twist off. If it had been tempered by the reverse process of heating the blade in the flame until a straw tint appeared and then allowing to cool, the blade itself would have cut quite as well, but the drill stem just behind it would still be hard and liable to break off and leave the drill blade in the half-drilled hole, a contingency to be avoided, as nothing is more annoying.

Take a clock screw. New screws are made from soft steel, and if used soft, the threads are liable to damage and the slots get burred and cut by the screwdriver. Therefore it is desirable to harden and temper them to resist wear and give the maximum toughness. A piece of binding wire is twisted round it, and it is held in a gas flame until red-hot, then quickly plunged in water to harden, tested and dried. Its head is then brightened on the top to show the colour and it is placed upon a " blueing-slip." This is a slip of thin sheet brass upon which small steel articles are laid to heat them for tempering. It has several holes of various sizes through which screws, etc., can be put with their heads resting on the slip. Placed in a hole in the blueing-slip, the slip is held over a spirit-lamp flame until a deep blue colour appears on the screw-head. The screw is then instantly tipped off the slip and allowed to cool.

The brightness and beauty of the blue surface will depend wholly upon the smoothness and degree of polish of the steel. If the screw will be visible and is desired to look nice, it must be smoothed and polished while hard, or else first tempered as just described, then the head filed or turned to shape and size, smoothed and polished and re-blued, finally to give a nice finish and appearance and also a protection from rust. A blue surface does not rust as readily as a bright one. Similarly with a clock hand. It should be smoothed and polished and then laid upon a blueing slip and "blued." Blueing can be done without previous hardening and is so done in the case of hands and some other parts, for appearance merely, but the blue is not so good or lasting, and of course the metal itself is still quite soft. Blueing soft steel does not harden it.

Again take a pair of pallets for a skeleton clock. Here the object is to harden them glass-hard to resist wear and take a high polish. Brittleness is no special drawback. They are therefore held by a piece of binding wire in a gas flame until red-hot, then plunged in water and smoothed with emery buffs and polished, no tempering whatever being done.

Hardening Brass.—Brass is one of the softer metals, an alloy of copper and zinc in the proportion of about two to one. Brass rod and sheet and brass castings as bought from the metal merchants are soft, and for many purposes are used in this condition. The castings especially are soft. Rod and plate being rolled or drawn are harder, for brass is hardened by pressure closing its particles together more tightly. The pressure may be applied by hammering, rolling between steel rollers under heavy pressure, or by drawing through draw-plates as in the case of thin rod and wire. Thus rolled sheet brass for clock plates is further hardened by hammering. Wire can be made hard and elastic for making special springs, etc., by drawing down smaller through draw-plates. A small piece of wire to make a clock pin or for some similar part can be hardened by hammering or by twisting it, which exerts pressure by a tightening-up process. Very thin brass may be hardened by burnishing it on a hard steel surface. A small piece of sheet brass for making a spring like that under the cannon pinion of the skeleton clock is hardened by hammering, then bent and filed to shape.

If it is required to soften a piece of hard springy brass, as when a piece of wire is to be hammered flat, or drawn or rolled out, it is heated nearly to a dull red and allowed to cool.

Thus, if a piece of thick sheet brass is to be hammered out thin, it is first softened, then spread with the hammer until it becomes too hard to spread further, softened again and re-hammered, and so on until it is of the required thinness. So with wire that is to be drawn many sizes smaller. After each time of drawing through the plate it is softened afresh. If this were neglected, it would become hard and brittle and incapable of being further drawn.

Gold, Silver and Copper.—These metals, used occasionally in clocks and their dials and cases, are all worked by the same methods as brass, being hardened by hammering, drawing, etc., and softened by heat. Gold especially is capable of being softened and hammered out to extreme thinness, possessing great ductility, and can also be made very hard and elastic by hammering. It is a good material for thin springs, even the hairsprings of watches have been sometimes made of it.

Soft-soldering.—This is a convenient method of uniting metals, but has no great strength. Its principal advantage lies in the ease with which it can be done and undone when required and the low degree of heat necessary, which is less than that of blue-tempered steel. Hence it follows that spring steel can be soldered without softening it and destroying its elasticity. By soft solder is meant tinman's solder, mainly composed of lead and tin. The process is simple and consists of brightening and cleaning the parts to be united, by filing or scraping and cleaning from all traces of grease by petrol. Then the surfaces are " tinned," *i.e.* coated with a thin layer of solder by applying soldering acid as a flux and heating until a little solder placed upon them "flows" over the surface. It is assisted in flowing by dipping a copper wire in the acid and spreading the solder over the surface until it "takes" nicely. Then more acid flux is applied, the parts brought together and heated until the "tinning" flows and then pressed well together. Finally a good wash in hot water is given to wash off the remains of the acid and prevent rust or corrosion.

Steel, brass, copper, silver, gold, etc., are all thus united easily.

In clocks the process is often used for repairs and is handy in that they are not injured by heat or riveting, etc. The rating screw of a pendulum rod is often soft-soldered into the rod. The barrel of a clock is sometimes made by soldering a piece of tube to the main wheel. The brass collets upon which wheels are mounted are usually soldered upon the pinion arbors, etc.

The soldering fluid used as a flux is made by dissolving zinc cuttings in hydrochloric acid until the acid will dissolve no more. The acid is then termed "killed," but is by no means harmless. It is still highly corrosive and must be kept away from tools and clockwork in general. Also after soldering, all parts must be well washed in water, hot is best, to prevent rusting. Oiling is no good, washing is the only effective method.

Brazing.—This is a much stronger method of uniting metals and consists in heating to a bright red and running brass-spelter or silver-solder into the joint. But the fact of having to heat to a bright red or more prevents its use in many instances. When brass is brazed with brass-spelter, it is nearly as strong as if solid. Steel united with silver-solder is very strong also.

In brazing or "hard soldering," as it is sometimes termed, the parts to be united are filed clean and if possible held together by pinning or binding with iron binding wire. A paste of borax and water is used as a flux and applied to the surfaces and to the pieces of solder. The grains of solder are then laid upon the joint, and a blowpipe and gas jet used to heat all together until the solder runs into the joint. The heat is applied gently at first to boil up the borax and not disturb the solder too much, then steadily and without pause until the solder flows. The secret is to heat both surfaces equally, as the solder will flow upon the hottest spot. So when soldering two parts, one of which is thicker and the other thin, the thick part must have the largest share of the flame, to ensure that both reach the running heat of the solder simultaneously.

After brazing brass, the parts may be rehardened by hammering, etc., and filed up clean. Steel parts after brazing may with care be re-hardened and tempered, but it is a little risky.

Brass spelter is simply low quality brass, containing rather more zinc than usual and runs at a slightly lower heat than good brass, thus enabling brass to be brazed. Silver solder is low quality silver—that is, silver and brass. For steel, standard silver, *i.e.* old spoons, parts of watch cases, etc., may be used for hard-soldering and makes good hard joints.

Cementing.—This is used to a small extent in clockwork for parts where no great strength and tenacity are required and where heat would be injurious. Shellac is used as a cement and the parts to be united are cleaned, warmed until a flake of shellac will run upon them, and brought into contact and allowed to cool gradually. The heat required is a little greater than that of boiling water, say about 250°, so it hurts nothing; even soft-soldered parts can be cemented with shellac. It is used principally for cementing pallet stones into their holes or recesses.

The chief thing to avoid in cementing with shellac is over-heating. The shellac must only be warmed until it runs nicely, and not heated until it boils, or it becomes brittle and loses its adhesive power. In re-cementing parts that have given way, all the old shellac should be removed and new applied.

French Silvering.—This process is used in clockwork for dials, etc. It has no great wearing properties, but yields a nice white silver surface and only consumes a microscopic quantity of silver. Protected by a layer of thin lacquer, silvered surfaces last many years and show up black figures and lettering very well.

The process is to clean and prepare the surface of the brass,

D

then rub on the silver in the form of a paste, and well wash, dry and lacquer.

The paste is made thus :—dissolve some nitrate of silver in water and add salt to precipitate the silver to the bottom. This is chloride of silver and should be separated and washed and allowed to settle. The water is poured off and the mass mixed with salt and cream of tartar into a paste. The dial is cleaned with bath brick and water and rinsed, then rubbed with salt and the paste rubbed all over with cross and circular strokes, using a pad of clean rag, until a white surface is produced. The dial is then well washed, dried, and lacquered. The appearance will depend a great deal upon the quality of the surface of the brass. The brass should not be polished, but given an even grain. After bath brick, pumice powder may be used, rubbed in one direction only, straight across ; this puts on a grain all in one direction which looks well when silvered. The very finest emery cloth will also give a good even grain.

Grinding and Polishing.—The usual cutting materials in use are emery, pumice, bath brick, rottenstone, oilstone dust and crocus or red stuff. Emery will cut hard steel and is used principally in the form of buffs, that is, flat wood sticks covered with emery paper. It can also be used for flatting plates in the form of powder mixed with water and used on a brush or a wood rubber. Pumice is generally used for polishing brass, as a powder, mixed with water and applied with a rag pad. Bath-brick is used in the same manner on a rag or brush, or sometimes, to flat a plate, it is rubbed on in a solid block, with water. Rottenstone is mixed in a paste with oil and used on a brush or rag.

Oilstone-dust and red stuff are used to polish steel, and are mixed with oil into a paste and used on soft steel polishers. Oilstone-dust cuts quickly and smooths out filing and turning marks, leaving a fine grey surface. Red-stuff applied to this grey surface produces that high polish seen on the steel work and pivots of good clocks.

Tools.—Fig. 24 shows a "broach" for opening out holes to any required size, clean and round. They are 5-sided hard-steel tapered rimers and are used with oil, by turning

round and round, or backwards and forwards in a hole already
drilled to enlarge it, or to make it smooth inside.   They may
be purchased from the thickness of a hair (a watch "pivot
broach") to $\frac{1}{4}$ in. or more in diameter for opening the barrel
holes of large clocks.   It is the practice in clock-making not to
drill a hole quite as large as required, but to use a drill a
shade smaller than the hole wanted, and " broach " it to size.

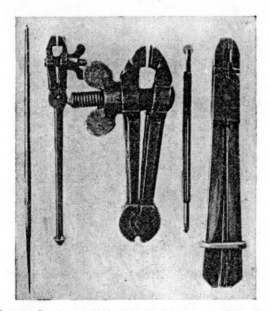

FIGS. 24.—Broach ; 25.—Pinvice; 26.—Handvice; 27.—Round Chamfering
Tool ; 28.—Sliding Tongs.

Broaching smooths and hardens the inner surface of the hole,
and as it is broached, any want of uprightness, etc., can be
corrected by bearing upon the broach more in the direction
required.

Fig. 27 is a round " chamfering " tool and is used to just
take off the burr or sharp edge from a broached hole.   It is
simply pressed upon the hole and turned with the fingers.   It
will also bevel the inner edge of a hole to take the riveting

when inserting a bush, or cut an " oil sink " around the pivot hole on the outside of the plate.

Fig. 25 shows a " pinvice." This is a small vice with a long handle for holding in the fingers. Its name indicates its principal use, which is to file pins or thin wire or rods. To file a pin tapered for inserting in a pillar, etc., the pin is rested in a hollow or groove in a piece of boxwood screwed into the bench vice and revolved with the fingers as the file goes across it. This keeps it round and true. The finger and thumb are placed on the stem handle, and as the file is pushed forward the pinvice is revolved forward to meet the file ; as the file is drawn back, the pinvice is revolved backwards also in opposition to the motion of the file. A little slow practice enables the trick to be acquired. This is usually the first lesson of a clockmaker's apprentice.

Fig. 26 shows an ordinary small " hand-vice," which, as its name implies, is a small vice between a pinvice and a bench vice in size, made to be held in the hand. It is useful to grip a piece of metal for filing, etc., or to hold a large broach so as to get more power than can be obtained by using the usual small wood handle. Also for holding a barrel abor when letting up or down a mainspring and many other similar purposes where a strong grip is required.

Fig. 28 is a pair of " sliding-tongs," a tool peculiar to watch and clockmakers. It is useful to hold a larger pin or rod than can be got in a pinvice, to hold small parts while filing, drilling, or broaching them, and to grip tight pins, etc., to draw them out. Its manner of use is evident from an inspection.

Drills and Drilling.—Clock drills are bought ready made in running sizes and with shanks of a standard size to fit the regulation drill-stocks and lathe chucks. Also twist drills can be bought for use in the lathe, and cut very clean round holes. Two methods of drilling are in use ; the old way by means of a " bow" which gives a to-and-fro motion, and by means of a lathe. The lathe drills more rapidly and cuts cleaner mainly because the motion is continuous and in one direction, causing the drill to cut a long shaving. The bow is handy in that it requires no special fitting up, the work being held up to the drill by hand, the back centre of the drill resting in a hollow in the side of the vice jaws. Fig. 29 shows this method. A drill is not a difficult thing to make and the old clockmakers always made their own. A piece of steel wire is taken a little smaller than the required drill. It is heated to redness at the end and hammered flat where the blade is to be. Then it may be filed up to the correct shape and the blade alone hardened. An emery buff will brighten it and it may be tempered to a pale straw colour to make it less brittle.

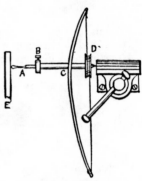

A few clockmakers use an archimedian drillstock, but the method is not a very satisfactory one.

For drilling brass a very sharp drill is required and good lubrication. For drilling steel, considerable pressure is needed but not such razor-like sharp-ness. Tempered steel is very

FIG. 29.—Drilling with a Bow.

difficult to drill and needs a sharp dead-hard drill, quite untempered and lubricated with turps. To drill soft metals like gold and silver, a very sharp drill must be used and water to lubricate. Every few turns the drill must be with-drawn, the blade cleared and the hole cleared out, or the drill will choke with the cuttings and break in the hole. The same remark applies to drilling hard wood such as ebony, or in drilling ivory, though no lubrication is necessary in drilling these substances.

Taps and Screwplates.—Making screws and cutting screwthreads is an important part of either clock making or clock repairing. It is true that screws are generally bought ready made, yet one has to be made sometimes, and it is frequently necessary to screw pins, etc., into holes, and all screws require threads cutting in the holes to receive them. This process is called "tapping" a hole, the thread being cut by a "tap." A tap is a piece of hardened and tempered steel with a screw thread cut upon it and having several flats or grooves giving it cutting edges. The hole is opened out by broaching

until the end of the tap will just enter.   The tap is then well
lubricated with oil and screwed into the hole with some force
and pressure.   Its cutting edges enable it to cut its way in as

it is turned round, and being slightly
tapered, it gradually cuts a clean
deep thread.   Fig. 30 shows a tap
as used in clockwork.

FIG. 30.—Triangular Tap.

To make a tap, a "screwplate" is required.   This is a
plate of hard steel having a series of graduated holes, cut with
a screwthread.   Fig. 31 shows a screwplate with plain holes

FIG. 31.—Plain Screwplate.

for cutting threads up to $\frac{1}{16}$ in diameter.   For larger screws a
different kind of plate is used, in which each hole has inside
cutting edges, as in Fig. 32.   The pattern of plate shown in
Fig. 31 simply forces a thread upon the screw by burnishing
and pressing into the hollows and forcing up the threads.
Such a method is only suitable for small work.   For anything

FIG. 32.—" Cutting " Screwplate.

larger it is necessary to actually cut the threads out, and the
pattern of plate shown in Fig. 32 does this, the two holes at the
sides receiving the cuttings.

A short length of tool steel rod is taken of a size a trifle
larger than the tap required.   It is first thoroughly softened by
annealing.   Everything that is going to have a screwthread cut
upon it must be made as soft as possible or the threads in the
holes of the plate will be blunted and spoiled.   The steel is
then held in a pinvice or hand-vice according to its size and
slightly tapered by filing until it will just enter the hole in the

screwplate. It is then well lubricated with oil and forced in, gradually using some pressure, but being careful not to over-strain it and twist it off. As soon as it goes hard and shows signs of binding, ease it back a half a turn, re-oil and turn forward again, proceed thus, easing back half a turn and forcing forward a turn, and so on until a good deep thread of full size is cut for a sufficient length. If the steel will not tap far enough on account of its thickness, withdraw it and reduce a little by filing and proceed again. When cut far enough, file it down triangular, like Fig. 30, by forming three flats upon it. Let them come nearly to knife edges at the extreme end and extend up eight or ten threads. Then pass the tap lightly through the plate again to remove burrs, and heat it in a gas flame to a good hardening heat and quench in oil. Hardening in oil produces greater toughness than using water, and toughness in a tap is essential.

When hardened, brighten the flats with a fine emery buff or on an oilstone, and temper it to a full straw colour. After tempering, smooth all three flats on the oilstone to give good cutting edges, file two flats upon the handle end so that it may be gripped firmly, and file upon it the number corresponding to the hole in the screwplate by which it was cut. Thus if No. 6, file six notches, and so on. Always use taps gently, keep them clear of cuttings, and well lubricated, ease them back every turn or so, and keep them sharp. Never screw a tap in its hole in the plate or both may be spoiled, and never try to tap a hole in metal that is not thoroughly soft.

Taps are made of many shapes and patterns. Some are square, some have two flats, some three grooves and so on, but for clockwork in general triangular taps cut best and are strongly recommended.

If it is desired to make a screwplate of a special size and thread and a good steel screw or a tap is available, it may be easily done. A steel screw may have its head filed square for gripping and its threaded part or " tap " filed triangular and slightly tapered. Then it may be hardened and tempered as a tap. For the plate, take a worn-out file. Soften it thoroughly and drill a hole in its centre, broach it to the correct size and with the greatest care tap it with the makeshift tap. Lubricate well and use plenty of patience. Tap it from both sides so as to get as full a thread as possible. If a large size, drill two other holes, one on either side, as in Fig. 32, and file through them with the smallest square file and see there are no burrs left. Then harden the plate and leave dead-hard if plain, temper to the palest straw tint (hardly discernible) if provided with cutting edges.

Old files make splendid screwplates and often outlast the shop-made kind.

Turning.—For the turning required in clockwork there is of course nothing so good as a proper " clock lathe." This is usually a small reproduction of an engineer's lathe, with a bed about 2 ft. long and 3 in. centres. It may be screw-cutting or not, it may have a slide-rest or not, but it must have good true fire female centres for turning pinions and pivots and a set of chucks for holding rods, arbors, screws and drills, etc. Also a face-plate for wheels.

Split chucks for holding rods are made in sets, of the

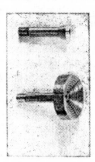

FIG. 33.—Split Chuck for Rods.

FIG. 34.—Split Step-chuck for Wheels, etc.

screwing-in variety like Fig. 33. Step chucks for holding discs or wheels are also made in sets, like Fig. 34. These resemble those used in watch lathes, but are larger.

A watch lathe is very useful for turning pinions and pivots, and for drilling the smaller holes in clockwork, but has not power enough for wheels, barrels, etc., or for drilling holes much over $\frac{1}{16}$ in. diameter. Also it is not large enough to swing parts like a crutch, or to take a main wheel like that of a grandfather clock.

For those who do much clockwork, a proper clock lathe is the thing to use. Any small 3-in. centre lathe is the next best, though as a rule these have only *male* turning centres and if used for clockwork must be provided with female ones. If the centre of the carrier chuck is softened and a small flat filed on its end, an accurate turning centre can be turned by a sharp graver point as it runs in the lathe, and deepened with a drill. The centre can then be re-hardened and tempered to a straw colour, and the result will be a female centre. The drilled hole must only be small, not so large as the smallest clock pivot, or they will stick in it and break off. Fig. 35 shows the altered centre and also an enlarged section of the point, showing drilled hole and coned opening.

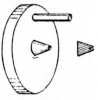

FIG. 35.—Carrier Chuck with Female Centre.

Similarly the back centre can be altered so that a clock wheel can be run between them for turning its pivots, etc. The lathe will then be to all intents and purposes a clock lathe, and the sets of special chucks, though very convenient, may be done without and more or less makeshift methods of holding parts adopted. A face plate and slide rest are a very great convenience and will do work almost impossible without them. For any one going in for actual clock *making* and taking a pride in their work, a screw-cutting lathe will be very useful.

FIG. 36.—Carrier.

FIG. 37.—Graver in Handle.

In it the spiral grooves of barrels can be cut, fusees, screws of all kinds, and nuts may be made true and perfect.

As an example of fine clock-turning, the method of turning a pivot will be described. Suppose the pinion to be truly centred in the lathe and to have the usual carrier affix (Fig. 36). A graver like Fig. 37 is used and the pivot turned down to the required length and diameter. A pivot should be straight and its shoulder square, with no root in it. Fig. 38 shows a properly shaped pivot at A and an imperfect one at B, with no squareness about the shoulder and a "root" like a lighthouse.

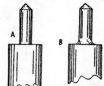

FIG. 38.—Shape of Pivots.

To cut the shoulder clean and square, the graver is held as in Fig. 39, while to turn the pivot down to size and smooth, the graver is held as in Fig. 40, on its side and cutting done by one of the bottom edges. A graver held thus cuts much better and more quickly than when cutting is done by the top facet

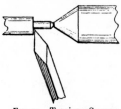

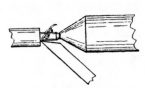

FIG. 39.—Turning a Square Pivot Shoulder.

FIG. 40.—Turning a Pivot.

while the graver is balanced upon its bottom edge upon the T-rest. On its side, the graver lies flat and firm upon the rest and is at a convenient angle for turning a straight pivot. When a graver is in good condition, a continuous shaving will be cut as in Fig. 40.

To turn tempered steel a sharp graver is required and it must be continually re-sharpened. If it is allowed to become dull it will burnish the steel instead of cutting in, and produce a glass-like surface that cannot be cut even by a sharp graver. When the pivot through turning with a dull graver becomes glazed in this manner, a fine file should be held upon it a moment to cut the glaze.

When turned to size quite straight and smooth, a fine pivot file followed by an oiled burnisher will give a finish good enough for rough and common clocks. But if the pivot is for a nice clock, it should be smoothed and polished, instead of filed and burnished.

This is done with a soft steel polisher. A strip of soft steel about $\frac{3}{16}$ in. wide and $\frac{1}{16}$ in. thick by 6 in. long will make a good polisher. It is slightly bevelled on the left-hand edge that lies next the shoulder of the pivot, and its surface has a fine grain put upon it by a smooth file to enable it to hold the cutting medium. For smoothing this use oilstone dust. This can be bought at the material dealer's and is mixed on a hard steel "stake," into a paste with oil, clock oil is best. The

"stake" is a slab of polished hard steel, kept in a wood or
metal box with a cover to keep out dust. The polisher is
rubbed upon the stake to charge it with oilstone. dust, and
then used upon the pivot after the
manner of a file, a somewhat rapid
motion being used and a light
pressure. Fig. 41 shows a section
of the steel polisher in position
upon a pivot to smooth it and
shows the bevelled edge next to the pivot shoulder.

FIG. 41.—Polishing a Pivot.

This will cut rapidly at first and then run smooth, doing
very little more work. Then wipe the polisher clean, re-file its
surface and re-charge it and apply again. About two or three
"wets" will smooth a fairly well-turned pivot and remove the
last trace of the turning marks. The last "wet" should be
worked down quite smooth till it ceases to cut at all. This
fines the surface and prepares it for polishing.

To polish, remove the pinion from the lathe and well clean
it from oilstone dust, as a grain or two of this will effectually
prevent polishing. Clean the lathe back centre; clean and
re-file the polisher, and then replace the pinion in the lathe
and charge the polisher with red-stuff and oil. This is mixed
upon another stake and kept scrupulously clean. Any dust or
grit in it will cause scratches and prevent polishing. Red-
stuff should be mixed thick and only a very little used upon the
polisher, just enough to charge its surface. Two or three
"wets" of this, the last one well-worked down, will produce a
glass-like polish. Finally a well-oiled burnisher can be used to
harden the surface.

Burnishing a smooth polished steel pivot improves it by
forming a very hard skin that resists wear. But burnishing a
filed pivot, though it makes it shine, forms a false surface. The
raised portions are merely spread out laterally to cover
the indentations and a surface of ripples is formed honey-
combed underneath. Such a surface, besides not being level,
will not wear well. In fact, burnishing is a poor substitute
for polishing, but a useful addition to it. A burnished surface
shows in its true light when it is required to "blue" the part
so treated. A screw-head, for example, if burnished and
"blued," will not show a true colour, but a polished head or
one merely left from an emery buff, will take a good colour.

Turning pinions and pivots is all done with hand gravers and the T-rest, but slide-rest work is useful in many clock parts. For turning a barrel arbor or a barrel, facing wheels, turning out their centres, making pillars, etc., the slide rest does better work and in less time than hand work.

For clock lathes a "cutter bar" is very useful, as shown in Fig. 42. This is a square bar as large as the slide rest will

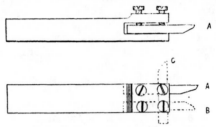

FIG. 42.—Cutter-bar for Lathe Slide-rest.

take, with a groove down each side and across the end in which cutters can be held. The advantage of this is that the cutter bar is firm and solid, and the cutters themselves may be quite small and made from $\frac{3}{16}$ in. square steel rod in lengths of 1 in. to $1\frac{1}{2}$ in. Such small cutters are easily filed up to shape and hardened and sharpened, and the bar allows them to be held in a variety of positions, across the end as at C, or down either side for right or left cutting as at A and B.

Figs. 43 to 48 show cutters for various purposes. Fig. 43 shows a pair of ordinary turning cutters, right and left. Their use is shown in Fig. 49. Fig. 44 is a round-nosed turning tool

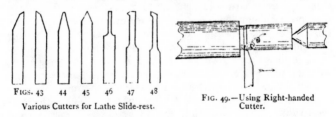

FIGS. 43   44   45   46   47   48
Various Cutters for Lathe Slide-rest.

FIG. 49.—Using Right-handed Cutter.

for reducing the diameter of a rod and will cut in either direction. So a cut may be made by traversing the rest along the rod to the end, then deepened and traversed back again. Fig. 45 is a screwthread cutting tool for V threads. Fig. 46 is a " parting " cutter. Its use is shown in Fig. 50, where it is shown cutting off a portion of a rod. Fig. 47 is an inside screw-cutting tool and Fig. 48 an inside turning tool for

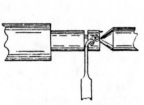

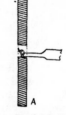

FIG. 50.—Using Parting Cutter.   FIG. 51.—Turning inside a hole.

turning inside a small hole, as, for instance, in turning the centre hole of a main wheel as shown in Fig. 51, A being a section of the main wheel held upon a face plate.

The screwing tools, Figs. 45 and 47, are of course only useful in a screw-cutting lathe.

A pattern of lathe very useful for clockwork is that which has a long leading-screw under the bed, which can be operated from the right-hand end of the lathe by a handle and thus make the bed the bottom slide of the rest. Its advantage is that a long rod can be turned from end to end without moving the rest, and the top slide can be used for taper cutting when necessary without interfering with the ability of the rest to travel straight along the bed and cut parallel. One of the first small lathes to give this advantage was the " Pittler," and now it may be found in many makes.

Filing.—A file is as much a cutting tool as a graver. Fig. 52 shows an enlarged view of file teeth. They are like saw teeth, or a succession of chisels one following the other. The idea is prevalent among some that a file is a thing to be rubbed to and fro over a piece of metal

FIG. 52.—Shape of File Teeth.

to wear it away after the manner of a piece of emery cloth. But a file needs careful and intelligent using like any other

cutting tool.   It only cuts on the forward stroke.   Consequently, all the attention should be directed to holding it
level and keeping an even pressure during the forward stroke,
making sure that it "bites."   The back stroke should be
quite light, the file barely touching the work, but just being
kept in touch so that the level of the next forward stroke is
not disturbed.

A new sharp file should not be used on steel if it can be
avoided.   The teeth are apt to break off.   Brass, on the other
hand, requires a new sharp file, or it cannot be properly cut at
all.   A file for steel should be first used upon brass until its
newness is worn off and it begins to cut a little more slowly.
In this condition it is at its best for steel.   For filing tempered
steel, when it is absolutely necessary to do so, use a file still
more worn, one that begins to cut ordinary steel slowly.   In
buying a new file for general use, and which has to be used
at once on all kinds of work, it is a good plan to chalk one
side a little to distinguish it from the other, and use the
chalked side upon brass only, the other being reserved for steel.
At least one sharp side will then be always available for
brass.

When files are finally worn out, they are useful for a variety
of purposes.   They make good lathe hand-cutters, special
screwplates, drawplates, etc.   Or the steel can simply be used for
various purposes of clockmaking, such as pallets, etc.   Small
round and square files make punches, chamfering tools, etc.

CHAPTER V.

OTHER PENDULUM TIMEPIECES.

THE skeleton clock described in Chap. I. is a "timepiece," that is to say, it merely shows the time, and does not either strike or chime the hours or quarters. These latter are "striking" and "chiming" clocks respectively as distinguished from simple "timepieces."

Eight-day English Dials.—Nearly allied to the skeleton clock, which is essentially an English form of clock, is the eight-day English "dial" clock. These clocks have barrels, fusees, train wheels and pendulums exactly like the skeleton clock, but the frame is composed of two square plates, and the dial is usually of sheet iron painted white with black figures, and about 12 in. in diameter. The cases of these clocks are generally of oak or mahogany, and the movement is contained and covered by a box back. Sometimes the pendulum is short so as not to project below the circle of the dial, when the clock is known as an eight-day "round" dial. At other times the pendulum is longer and comes about 6 in. below the dial edge, the box-back being extended downwards to contain it. It is then known as a "drop" dial.

To clean one of these clocks, first take off the pendulum. To get at it there is a door at the bottom of the case and another at the side where the hand can be inserted to get at the suspension. Then remove the hands, and laying the clock on its face, withdraw the wooden pegs that hold on the box-back of the case, and remove the back. Draw out the pins that hold the dial feet to the front plate and the movement can be lifted off complete, leaving the dial affixed to the case front. To remove the movement with its dial, by taking out the little screws that hold it to the wood circle, is unwise; as these screws are small and the movement heavy, the less they are disturbed the better. They so easily overturn and get loose.

The movement is taken to pieces, cleaned and repaired exactly as the skeleton clock in every particular, but not being visible it does not require such careful handling or such perfect polishing of the plates.

These clocks go about as well as the skeleton clock and make excellent kitchen, shop, or public-room clocks. Fig. 53 shows one.

American and German Dials.
—Very much like them in outward appearance are American and German eight-day dials, but the movements differ very much.

FIG. 53.—Eight-day English Dial Clock.

To take one apart, unhook the pendulum bob, and take off the hands. The minute hand is only held by a pin and collet. The hour hand is pushed on friction-tight and needs drawing off with cutting nippers or some other tool that will just go under its edges and give a purchase. Next remove the dial, either by taking out the little screws around its edge, or by withdrawing the wooden pins that hold on the case front. Some are made in one way and some in the other.

The movement will then be found screwed to the case back by three or four screws, or held in wood clamps. Undo these and lift the movement out. It is usually quite a small and lightly made affair. The plates are of thin brass with openings stamped out of them to save weight and enable the parts to be seen. The mainspring is not contained in a barrel, but the outer end is generally looped around the lower corner pillar of the frame and the inner end hooked directly to the axis of the main wheel. As the mainspring is strong and opens out to a large size when unwound, it is advisable before taking the clock apart, to confine it within limits for the time being. This is done by winding it up close and slipping on a "clamp" (Fig. 54). This is a piece of stiff iron wire rather more than a half-circle. New mainsprings when purchased are wound up and confined in them, and one or two various sizes should always be kept at hand for use when taking clocks apart.

The clamp should be held in position while the pallets are removed and the train is allowed to run. As it does so, the mainspring gradually unwinds and expanding, fills the clamp. It can then unwind no further and the train comes to rest. The spring is then harmless, and the pillar pins or nuts can be removed and the movement taken apart. But before doing so, the pendulum spring and wire should be removed. Generally the spring passes through a slit in a brass stud in the front plate. This slit can be opened with a pocket knife and the spring taken out. These clocks vary very much in the method of suspending the pendulum, but with the clock open to inspection, whatever method is used, no difficulty will be found in detaching it.

Fig. 54.—Mainspring Clamp.

An examination of the clock will reveal many points of difference from the English pattern. The wheels are thin and stamped from sheet metal. The teeth, however, are machine cut and generally clean and well shaped. The pinions are of the "lantern" pattern (see Fig. 102, p. 130) consisting of a circle of round wires. This pattern of pinion has many advantages and is well adapted for running in a dirty and dusty condition, as the dirt pushes through and does not stick between the leaves. They are also cheap to make.

The escapement is generally inverted, that is, the pallets are below the wheel, or sometimes at its side, and consist of a strip of steel bent to the form required, and hardened and tempered. This is also for ease and cheapness of making and answers very well.

The result is a clock costing less than half as much as the English pattern and half the weight; but also having less than half the wearing qualities and not to be compared for steady timekeeping. Still their price and appearance causes very large numbers to be found in use, both in timepieces and strikers, and they are quite a stock feature of the clock-repairing shop.

To clean such a clock, place all the parts in petrol in a

E

basin and rinse and brush them clean, wipe them on a duster, and when dry peg out the holes clean, all as described in Chap. I. Clean out the pinions between the wires well, and look carefully over the wheels and pinions for bent teeth or wires. The wheels being thin are especially liable to such injuries, and any shock caused by a key slipping during winding, a mainspring breaking or a click giving way often results in bent teeth or pinion wires.

The mainwheel and mainspring attached alone should not be put in the petrol, but should be wiped clean with a duster. When confined in a clamp as above described the mainspring outer loop can be slipped off the pillar to which it was affixed, leaving the plate free and clear to put in the petrol, but if no clamp happened to be available with which to confine it, other means must be adopted, such as winding up and tying round with string or copper wire. Then as the string or wire is bound to be passed round the pillar as well as the spring, the mainwheel and spring cannot be detached from the plate, and the plate cannot be put in the petrol. Still parts of it can be cleaned by a brush with a little petrol upon it and the rest wiped clean as well as possible.

If one of these clocks is taken apart without first confining the spring, very great difficulty will be experienced in getting it together again, as the stiff spring opens out to such a large diameter that the wheels can scarcely be forced into their places beside it without damaging their teeth or pivots.

Before putting together again, the centre arbor should be taken in hand and the stiffness tested. To set the hands, the centre arbor is made to turn in the wheel friction-tight, generally being pinned up tight against a spiral spring on the arbor and a washer and pin, or sometimes by means of a flat brass spring against the face of the centre wheel. It is annoying to find this too easy after the clock has been put together and started going, as it is impossible to tighten it without taking the clock apart again. If left easy, the clock continues to go, but the hands lag behind.

If too easy, unpin it, and if a spiral spring, extend it by drawing the coils apart to give it more spring. If a flat spring, curve its ends a trifle more to give more grip. When all is right, place the wheels in position, put on the top plate and pin down one pillar, then exerting pressure upon it, get in the wheels one by one until the plate goes on. Often the top escape wheel cock will require springing up to take out the escape wheel or to put it back again, but this does not hurt it.

There is one good feature in the escapements of these clocks, the pallet depth is adjustable by a simple means. Fig. 55 shows the escapement. The pallets work upon a steel pin fixed in a small arm, A. This arm is riveted to the front plate at one end and can be turned stiffly with a pair of pliers to or from the escape wheel to regulate the depth. The pallets are generally of the recoil form and act exactly as those in Fig. 23, and the depth should be adjusted so that it is as deep as possible without catching the teeth points. When the pallets are badly worn, the arms of the escape wheel must be bent up or down to make the wheel cup-shaped and bring the action to a fresh unworn part of the pallet faces.

FIG. 55.—American Recoil Escapement.

When wheel teeth are broken they are repaired as in Fig. 17. Worn or damaged pinion wires must be replaced by new ones. They are held in by the brass being burred over their ends, and can be easily forced out with pliers. In fitting new ones, use wire of exactly the same size, needles do well, and insert and burr the brass over them again as before. If merely pushed in tight they will work loose. Each wire should be cut a little too short, so as to go well in and the brass forced over their ends to prevent them coming out.

Pivot holes are bushed as Fig. 21 by the usual bushing wire. New mainsprings are bought ready to put on in place of the old ones and will need no special directions, except that the wire clamp should be left on until the clock is put together and wound up.

The clock having been put together, the pallets placed in position and the depth adjusted, the mainspring may be wound up and the clamp taken from its coils. If in good order, it should trip easily. A little clock oil should be placed upon each pivot and the pallet faces, and also on the coils of the mainspring. The movement can then be replaced in its case, and the pendulum spring and wire put back and made fast. The slit in the stud which holds the suspension spring can be closed by nipping up with cutting nippers. This done, hang the clock up and hook the pendulum bob on. When started thus without the dial and hands, any little faults can be easily seen and corrected. See first that the case hangs perfectly upright, and then adjust the beat of the clock so that it is perfectly even. This is done by bending the wire crutch with pliers, at the same time seeing that the pendulum rod hangs centrally in the crutch loop and does not touch either the front or back of it. The shake of the rod in the loop should only be slight, but it must not be so tight as to stick. Perfect freedom is essential, as there is constantly a little movement here owing to the fact that the centre of motion of the pallets is above that of the pendulum, and for this reason, a small drop of clock oil should be placed in the crutch loop. The crutch loop must not project too far forward or it may just touch the dial plate and cause stoppage.

It is as well to let the clock go a few minutes like this to make sure that all is right, for owing to the shake of parts and the occasional unevenness of the escape wheel teeth, one cannot always be certain that an odd tooth will not catch the pallet points. After five minutes the dial and hands may be put on and the clock completed. A slightly oily rag rubbed over the dial will clean it and a polish with a dry duster to finish. An oily rag is also a good thing to polish up the case.

The principal repairs to these clocks have been indicated above, but there are some others which may be mentioned. When pallets are finally worn out, new ones complete can be purchased for a few pence, the old ones being taken as a guide for size and shape. It will be noticed when the pallet points are placed against the wheel, that the distance between them is not equal to an even number of tooth-spaces, but consist of $4\frac{1}{2}$, $5\frac{1}{2}$, or $6\frac{1}{2}$ spaces as the case may be, and when pallets are lost and must be replaced, a pair should be selected that when placed against the wheel with one point in contact with a tooth point, the other pallet point comes just midway between two teeth, then a very little attention with an emery buff, or perhaps none at all, will be needed to adjust them in the clock.

A broken click is sometimes met with, and is fairly easy to replace. Fig. 56 shows the click. It is fastened to one of the main wheel arms or to its rim by a brass rivet. These clicks, also pillar, nuts and other small parts, can be purchased very cheaply and hardly pay to make. Still if one is not at hand, it is not the work of many minutes to make one from a fragment of old clock plate or a wheel rim. The rivet by which it is attached must be very firm and solid, and a little care taken in filing one up with a shoulder to enable it to be riveted tight without binding the click is well repaid. A rivet is shown at A.

FIG. 56.—Click and Stud.

For touching up dials, ordinary enamel paints are useful, as sold in small tins, white for the dial plates and black for the figures, but it is not expensive to have a dial properly repainted by a dial painter, the principal item is the carriage each way to the material dealers.

Thirty-hour American Timepieces.—These resemble the eight-day movements very closely, but are smaller and lighter and have one wheel less, and exactly the same directions may be followed for their cleaning and repair.

Eight-day French Timepieces.—Fig. 57 shows the movement of one of these as fitted with the ordinary black marble or wood case. The method of holding in the case is peculiar to these clocks and consists of two straps of brass, shown in the photo. The movement is put into an opening in the front of the case and two long screws are passed through a metal rim forming the case back and draw the side straps up tight. This method permits the movement to be turned round slightly to set accurately in beat, and then screwed tight.

The workmanship of these clocks is generally good and more like the English clocks first described. The wheels are substantial and the pinions nicely cut and polished. There is one great point of difference, however, and that is the mainspring and barrel. French clocks have "going barrels," that is, the main wheel teeth are cut round the barrel itself, which is thus barrel and main wheel combined. It is shown in the lower part of the photo, and is more distinct still in Fig. 71, a photo of a carriage clock. It is a simple method, doing away with fusee and chain, but has the disadvantage that when a mainspring slips or breaks, the shock is very liable to bend or break the barrel teeth, and these are troublesome to replace, there being so much pressure upon them. Also the power

FIG. 57.—French Timepiece Movement.

of the mainspring varies very much during the week, being greater when just wound up and less towards the end of the week. But in these clocks this is not so objectionable as in most others. The workmanship being good and the pendulum fairly heavy, the clock does not require much power to drive it; this enables a long, thin mainspring to be used, giving sufficient turns to keep the clock going a fortnight. Being wound every week, the first seven days do not show so much variation in power as they would if the spring were shorter. On the whole the timekeeping of these clocks is almost equal to that of eight-day English clocks with fusee.

To take apart, first unhook the pendulum, undo the strap screws and take off the case back, and draw the movement out from the front of the case. Remove the hands and take out the pins in the dial feet that hold the frame to the brass edge.

The pallets can be taken out and the clock allowed to run down after applying oil to the pivots. Or the mainspring may be "let down" with a key, holding the click back with the thumb of the hand that holds the movement. This is a little risky as if the click slips, the key may run back in the hand and injure the fingers as well as risking damage to the wheel teeth.

Then the click and ratchet can be removed, and the motion work taken off, the pins taken out of the pillars and the frame taken apart. The barrel cover can be prized off with a screw-driver and the arbor removed.

To clean, first polish up the wheels, etc., with "Globe" metal polish and place in the petrol. Strip the plates of all screwed-on parts, including the suspension of the pendulum, polish them up and wash in petrol. When dried off, polish up all parts with a watch brush and dry chalk as described in Chap. II., and peg out the holes quite clean. Peg the pinion leaves clean, wipe out the mainspring and barrel, polish up the barrel, and apply fresh clock oil to the coils of the spring. The barrel arbor may be replaced and the cover snapped on again. If the cover goes on hard, lay a duster over it and knock it in with a boxwood mallet or a wooden screwdriver handle. Oil the pivots of the arbor where it turns in the barrel. The wheels, etc., can then be replaced and the frame put together.

The pendulum suspension arrangement, shown well in Fig. 57, should all be taken apart, cleaned and polished. This is an ingenious arrangement, the regulation being accomplished from the dial of the clock with a watch key. By means of two small gear wheels a screw is revolved which shortens or lengthens the acting part of the pendulum suspension spring. This is designed so that approximately one complete turn of the square over the Fig. 12 on the dial makes a difference in time of one minute per day, and in regulating these clocks it is useful to remember this.

The suspension spring itself is delicate and the least kink ruins it, causing the pendulum to "wobble" and the clock to stop.

When together, oil all the usual parts, including the pivots, pallets, crutch, click and clickspring.

There are two patterns of motion work in these clocks. In the best, the cannon pinion is made like Fig. 10 (p. 13). These give the least trouble, but if too loose the cannon pinion may be removed and the spring portion bent inwards a trifle. This effectually tightens it. In this pattern of motion work, the minute wheel is held by a separate cock and its lower pivot must not be forgotten when oiling ; it is best to oil it before putting in place.

In the other pattern of motion work found in the cheaper clocks, the centre arbor revolves friction tight inside the centre wheel, much after the manner of the American clocks described earlier. These when loose are more trouble to tighten. To tighten one, take off the wheel, close the brass pipe that holds it, and put it on again. It is quite useless trying to tighten it when the clock is together, therefore it is advisable to carefully test it before placing in position in the frame.

When all is correct, put the clock in its case again and screw the strap screws lightly, place the clock on its shelf and listen to the beat, causing it to go with as little movement of the pendulum as possible. Adjust the beat by turning the movement round in its case ; when correct, remove the clock from the shelf and tighten up the strap screws finally.

French clocks being smaller and of finer workmanship than English or American clocks, require greater care and a more delicate touch to repair them, especially when we come to the smaller wheels and the escapement.

New mainsprings are put in after the manner of English clocks, but not being so strong are easier to handle (see Chap. III.). When very sticky with dry oil they should be taken out, cleaned and replaced, otherwise they need not be removed from their barrels.

Broken teeth in the barrel cannot be replaced in the same manner as in train wheels, there being not sufficient thickness

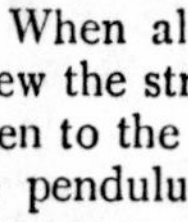

FIG. 58.—Replacing Broken Barrel Tooth.

of brass, but two holes should be drilled where the tooth should be, and two steel pins *screwed* in tight and filed up to shape as in Fig. 58. This two-pin arrangement is not of course an ideal tooth, but it gives a maximum of strength, and is the best that can be done.

Broken pivots are replaced by the method described in Chap. III. (Figs. 39, 40, and 41), a delicate touch and a very sharp graver being required for their turning. New pinions can be purchased ready hardened and tempered, with the leaves cut and polished. These require pivoting, and the wheels mounting upon them.

The larger holes in the plates are bushed as described in Chap. III. (Fig. 21), but for the smaller holes clock "bouchons" are better. Bouchons are little pieces of brass wire with the ends turned down, tapered and central holes drilled. A short distance from the ends they are partially cut through, see Fig. 59. To use one, open out the pivot hole until the bouchon goes halfway in, then drive it in and break it off. Level the end and drive home with a flat punch and finish off by smoothing and polishing on the inside surface, and cupping with a circular chamfering tool on the outside. Finally open it out to fit the pivot.

Fig. 60 shows the escapement. It is generally of the recoil variety like that of the skeleton clock, Fig. 23, but does

FIG. 59.—Clock "bouchon."   FIG. 60.—French Recoil Escapement.

not span so much of the wheel and has straight pallet faces instead of curved. Exactly the same remarks apply to it as to the other. To shift the pallets for wear is easy, as they are only driven on a tapered arbor, generally square, but do not attempt to drive them further when already tight, as being glass-hard they may easily break in two. First take them off and reduce the arbor a trifle, then push on again. If too loose they can be soft-soldered on, using a very little solder and well washing afterwards. If by chance a pair of pallets breaks through being driven on too tight, soft-solder them together again, pressing very close and using only a trace of solder, and they will generally do again. Use shellac to fasten a soft-soldered pair of pallets upon their arbor. To adjust the pallet depth, one of the pivots runs in a hole in an eccentric disc riveted in the plate as at A, Fig. 60. This eccentric has a screw-driver slot cut across it, by means of which it can be stiffly turned to cause the pallet to approach the wheel.

Other kinds of French Pendulum Clocks.—Circular

movements similar to Fig. 57 are sometimes fitted into brass drum cases about 4 in. diameter and provided with little short pendulums of about 3 in. in length, fixed direct to the pallet arbors and having no crutches or suspension springs. These are known as "tic-tacs" and are a very unsatisfactory kind of clock, fortunately no longer made and passing rapidly out of use. No special directions need be given concerning them.

Variations of the escapement are sometimes seen to make fancy clocks, with see-saws, swings, etc., instead of plain pendulums. The timekeeping of all these arrangements is poor. In the case of the swing, which is perhaps the com-monest, it is not necessarily so, but the form of escapement and crutch generally seen in connection with it precludes good timekeeping. To obtain a motion of the swing at right angles to that of the wheels, a double pin escape wheel is used acting on circular disc pallets and giving a circular motion to the crutch. This causes much friction, which could be avoided by the use of a crown wheel in the train. The difficulty of course is that the swing must travel backwards and forwards from the front to the back of the clock, while the train wheels revolve in a plane at right angles to that of the pendulum.

A figure holding a pendulum suspended from its hand, sometimes with the clock in the pendulum bob, and sometimes outside ; a globe with the hours painted around it, revolving behind a pointer, or surmounted by a conical pendulum (swinging in a circle and driven by a revolving finger), and other fancy arrangements, are sometimes seen. All these can be cleaned and repaired according to the directions here given for French timepieces, the only differences being in the escapements and in the manner of fixing the movement in its casing, etc.

Four-Hundred-Day Clocks.—There is, however, one

other kind of clock in this class that deserves special mention, and that is the four-hundred day clock provided with a heavy disc pendulum that vibrates something after the manner of a balance wheel. Whether it should be classed as a pendulum or a balance is a doubtful point.

Fig. 61 shows the pendulum or balance. It consists of a heavy circular brass plate or disc A, suspended centrally by a thin steel wire, D. When given a circular motion, it will spin until the tension on the suspending wire arrests it and causes it to return, the action being on the principle of the familiar roasting-jack. A small arm, E, is clamped to the wire near its top, and it is to this arm that the pallets give impulse. BB are two auxiliary circular weights which rest upon A, and by means of a right- and left-handed long screw C, the weights

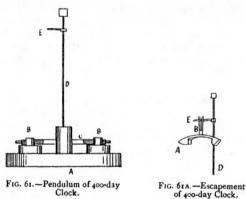

FIG. 61.—Pendulum of 400-day Clock.

FIG. 61A.—Escapement of 400-day Clock.

BB can be removed nearer the centre or nearer the edge for regulating.

Fig. 61A shows the escapement, or rather the pallets and the driving fork. A=the pallets of ordinary form. Above them the fork or crutch B extends and engages the arm E (see Fig. 61), attached to the suspension wire D, and gives it impulses to and fro. It is an ingenious arrangement and designed to give a slow motion and requires as little power as possible. To go four hundred days, the train consists of more wheels than usual. A going barrel, fairly large, drives a train of five wheels. The last is the escape wheel, and as the clock beats only ten per minute, this revolves in about eight minutes. Next comes the centre wheel, carrying the minute hand and motion work as usual, then three intermediate wheels and the barrel. With a ratio of wheels to pinions of about 10 to 1, this gives the necessary length of run.

Fig. 61 shows the simplest form of pendulum fitted to these clocks, but many patterns of " bob " or disc are used. Some have tubes of mercury for compensating for temperature errors. Others are of fancy design, resembling engine governors, etc., in form, but the principle is the same. Owing to its slow motion, the pendulum has no great governing power and consequently these clocks do not show a very even rate for the year.

To set in beat, the pendulum being brought to rest, the pallets should be level with one escape tooth just in the middle of an impulse. They can be so adjusted by turning the arm E upon the pendulum wire or bending the fork B (Fig. 61A).

They are repaired and cleaned exactly as ordinary French clocks, and being under glass shades, it will pay to nicely polish all parts.

Vienna Regulators.—A very satisfactory kind of pendulum timepiece for the wall is a Vienna Regulator. These clocks are carefully made, have dead-beat escapements, good long pendulums, and are driven by a weight. The wood rod of the pendulum makes for steady running, as the effects of changing temperature upon wood are not as great as upon metal.

The movement is much after the style of the French time-piece as far as workmanship and finish is concerned, but it is generally between plates of a square or oblong form, and the pendulum, instead of being suspended from a cock upon the back plate, is supported by a bracket fixed to the back of the case. This also conduces to steady timekeeping, as the suspension is firmer and the pendulum made independent of the rigidity or otherwise of the movement. There are several other refinements in these clocks, among them an arrangement to keep the clock going while being wound, by means of a "maintaining spring and ratchet" interposed between the main wheel and the barrel. The winding ratchet is upon the end of the barrel as usual, but instead of placing the click and clickspring upon the mainwheel, they are placed upon an intermediate ratchet-wheel with fine teeth. Affixed to the other side of the maintaining ratchet is a spring of curved form which presses upon the arms of the mainwheel and while the clock is going is always under tension, the clock being driven through this spring, it forming the only connection between the barrel and the mainwheel. When a key is placed upon the winding square and the barrel turned round to wind up the line, the maintaining ratchet is held from moving by a long-pointed "detent" engaging its teeth, and the maintaining spring, being kept under tension, continues to drive the clock.

Another refinement is an adjustable crutch for setting the clock in beat without bending the parts.

To clean one of these clocks, take apart as usual and polish and clean all the parts as a French clock. Take the barrel and maintaining work all to pieces, and in putting together see that it all acts freely and is well oiled at all points of friction.

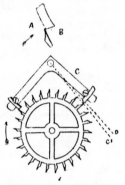

The escapement is a "dead-beat," that is, between each beat the escape wheel stands still, the teeth points resting upon the "dead" faces of the pallets, instead of recoiling. Fig. 62 shows the escapement. The pallet arms C are of brass and have circular slots cut in them in which the steel pallets lie, held by clamping screws which permit them to be moved for adjusting the depth. In the enlarged

FIG. 62.—Vienna Regulator Dead-beat Escapement.

view of one pallet and a tooth, the tooth point is shown resting upon the "dead" face A of a pallet. As this pallet moves outwards from the wheel, the tooth point escapes from the dead face and traverses the "impulse" face B, giving the pendulum an impulse. Another tooth then falls upon the dead face of the other pallet and so on. In adjusting the depth, see that the teeth points drop upon the "dead" faces and not upon the impulse faces of the pallets. This may be tested by placing a finger upon the crutch and leading it slowly across, watching carefully the exact spots upon which the teeth drop. Although the teeth *must* drop upon the dead faces A, they must drop as near to the corner as possible without missing it. This is important, as if the teeth drop too far up the dead faces it makes a large arc of motion of the pendulum necessary to keep the clock going, and the weight is not sufficient for this, consequently the clock will stop from insufficient power.

The escapement of every Vienna Regulator should be examined in this way and adjusted to "lock" on the dead faces of the pallets as lightly as possible, that is, to drop as near to the pallet corners as possible without missing them. As in French clocks, one pallet pivot generally runs in an eccentric like that shown in Fig. 60 at A. This is to adjust the *extent* of the drops and equalize them. It does not affect the *point* of the drop. That is adjusted by the steel pallets themselves and their clamping screws as explained above. But if on examining the escapement the drop upon the first or entering pallet is greater than that upon the exit pallet, the pallets must be lowered; or if that on the exit pallet is excessive, they must be raised, by means of the eccentric.

In putting the clock together, first hang up the pendulum, having seen that its groove for the crutch pin is clean, then slide the movement into the grooves of its supporting bracket, seeing that the crutch enters the pendulum slot. Care must be taken that the crutch is free in the pendulum. If the movement is pushed too far back, the crutch will bind against the pendulum rod. The centre of the crutch pin should work in the slot, there being a little freedom for the pendulum rod to work backwards or forwards without touching the base of the crutch pin.

To set in beat, start with as little motion of the pendulum as possible, and adjust the screws on the bottom of the crutch until the beat is perfectly even.

When in good order, the timekeeping of these clocks is excellent, and their wearing qualities good, owing mainly to the light and even pressure of the driving weight.

Holes are bushed, pivots and pinions repaired, etc., as before described. Wear on the pallets is easy to rectify. Simply buff out the marks, polish the impulse faces again, and replace the pallet and adjust it to depth properly by its clamping screw.

As usual, oil all pivots, pallet faces, crutch pin, and do not forget the weight pulley.

Other Forms of Dead-beat Escapements.—Any of the other clocks described may have dead-beat escapements similar in principle to that of the Vienna Regulator. Thus the English skeleton, dial, or other clocks may have the English form shown in Fig. 63. In this form the escape wheel-teeth are pointed and sloped forward, and the pallets are simply filed

FIG. 63.—English Dead-beat Escapement.

FIG. 64.—French Dead-beat Escapement.

out of solid steel and hardened and tempered, there being no real "adjustment" at all. In hardening these pallets, the centre is generally left soft, the ends only being hardened; this is to allow of wear being buffed out of the faces, and the pallet arms being bent closer together to adjust the locking again, a clumsy method at best, having nothing to recommend it. AA are the dead faces, and BB the impulse faces.

The common French form is shown in Fig. 64. As before, AA are the dead faces and BB the impulse faces. In this form also the pallets are made from one piece of steel and no adjustment is possible except by the usual eccentric in the plate to bring them nearer to the wheel.

Another form often seen in French clocks is the "pin pallet" shown in Fig. 65. In this form the fronts of the teeth are quite straight and the pallets slide up them during the "dead" portion

FIG. 65.—Pin-pallet Escapement.

of the movement. Sometimes the pins are steel, hard and polished, and sometimes of agate or onyx. Agate pallets hardly show wear at all and are to be preferred; they are made as at A (Fig. 65) from a round pin half cut away, and cemented in holes in the pallet arms with shellac. The holes being a little larger than the pallet pins, a little adjustment is possible by warming the shellac and moving the pins. To be correct, the teeth points should "drop" exactly upon the centres of the pins, and they may be shifted until this is so. By turning a pin round upon its own axis a little, a tooth may be made to drop off either earlier or later as desired. As in the Vienna form, bringing the whole pallets nearer to the wheel decreases the drop upon the entering pallet and increases that upo : the exit pallet. When set correctly the flat sides of the pin pallets should point to the escape wheel centre.

This escapement is troublesome to adjust unless one thoroughly understands its action, but when right is most excellent, and offers a cheap and easy solution of the "jewelled escapement" question. The form with steel pins is bad, as they wear quickly.

The English and French forms of dead-beat escapement shown in Figs. 63 and 64 are sometimes jewelled with agate

or other stones as shown in Fig. 66 by having a slot filed out and a block of stone cemented in, the stone being then ground

FIG. 66.—Jewelled Pallet.

and polished level with the steel surfaces. The pallets of regulators and other fine clocks are nearly always so treated.

Solid English recoil pallets like Fig. 23 may also be grooved and jewelled much to their advantage, but the writer has never seen any so treated.

American and German clocks frequently have dead-beat escapements like Fig. 67, AA being the dead faces, BB being the impulse faces.

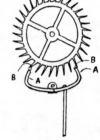

FIG. 67.—American Dead-beat Escapement.

In all clock dead-beat escapements, the dead faces of the pallets are portions of circles struck from the pallet pivot as a centre, so that as the pallets move, the teeth points resting on them remain stationary.

Variations of the forms of timepieces described in this chapter are sometimes met with, as, for instance, French Regulators which are French clocks in little glass cases, with usually a fancy pendulum and agate pin pallet escapement in front of the dial to look nice. Or English timepiece movements with a weight and barrel like a Vienna Regulator, and a long pendulum, in long wood cases and having large painted wood dials about 24 in. across. These are known as "act of parliament" clocks or "sheeps'-head clocks," but no special directions will be needed for their cleaning or repair. The directions for repair and the descriptions of escapements in this chapter apply to almost all clocks.

F

CHAPTER VI.

PENDULUMS.

Length of Pendulums.—The time of vibration of a pendulum depends entirely upon its length. Weight has nothing to do with it at all. Hence adding weight to a pendulum will not make the clock go slower unless the added weight is placed below the bob, when the effect is to lengthen the pendulum. Similarly if the added weight is placed above the centre of the bob, the clock will actually go faster, because the pendulum will be shortened, as measured from the point of suspension to the centre of the bob.

The exact length of a pendulum is from the point at which its suspension spring bends to the "centre of oscillation." This is variable according to the shape and make of the pendulum. The point at which the suspension spring bends is also variable, depending upon its stiffness. So in practice we measure a pendulum from the top of the suspension spring to the centre of the bob and the result is very near the truth.

The length of a pendulum beating exact seconds is, in England, 39·14 in. or about 39⅛ in. It varies from 39 in. at the Equator to 39¼ at the Poles. The force of gravity determines the rate of vibration of a pendulum, and that varies with the latitude, hence the differences.

The length of the seconds pendulum is a convenient starting-point from which to calculate the length of any other pendulum, or the time of vibration of a pendulum of any other length. It has been ascertained that the length of a pendulum varies as the square of the time of its vibration. Thus for a pendulum to take twice as long to vibrate, its length must be the square of 2 = 4 times as long. The length of a two-second pendulum would therefore be 39·14 × 4 = 156½ in. Or the length of a pendulum to vibrate twice as fast would be $\frac{1}{4}$ of the length, which gives the length of the $\frac{1}{2}$-seconds pendulum as $39·14 \div 4 = 9\frac{3}{4}$ in. or that of the $\frac{1}{4}$-seconds pendulum $= 2\frac{1}{2}$ in. Intermediate lengths give intermediate numbers, and clocks are found with pendulums of all lengths from $2\frac{1}{2}$ in. to the seconds pendulum, and in turret clocks sometimes up to two seconds or $156\frac{1}{2}$ in.

To calculate the length of a pendulum for any given number of vibrations a simple proportion sum is needed, using the seconds pendulum as a standard or basis of calculation. Thus to find the length of a pendulum beating 100 per minute, we state the sum thus :—

As $100^2 : 60^2 :: 39·14$ in. to the required length

or, $\dfrac{60 \times 60 \times 39·14}{100 \times 100} = 14·09$ in. $=$ the length required.

Or if we want to find the number of vibrations per minute of a pendulum 20 in. long, we state the sum thus :—

As 20 in. : $39·14 :: 60^2$: the square of the number required,

or $\dfrac{39·14 \times 60 \times 60}{20} = 7035$, the square root of which is 83.

Any other numbers or lengths may be substituted for those above and worked out by the same method.

The following short table gives the number of vibrations per minute, and the corresponding pendulum length, and intermediate numbers may be estimated nearly enough for many practical purposes.

Number of vibrations.	Length.	Number of vibrations.	Length.
30	156·5	120	9·75
60	39·14	130	8·3
70	28·75	140	7·2
80	22·0	150	6·25
90	17·4	160	5·5
100	14·1	185	4·0
110	11·6	240	2·5

The "centre of oscillation" of a pendulum is largely dependent upon its centre of gravity, therefore if its centre of gravity is high up, its centre oscillation is also high. From this it follows that for the centre of oscillation to be about the centre of the bob, the rod must be comparatively light and the bulk of the weight of the pendulum in the bob. This must be taken into account when measuring pendulums, and those with heavy rods and light bobs must be given a little extra over-all length.

Compensation Pendulums.—All metals expand under the influence of heat and contract in cold. The following short table gives the proportion in which they do so, approximately :—

Invar	= 1
Steel	= 14
Copper	= 21
Brass	= 25
Zinc	= 34
Lead	= 35
Mercury	= 200 (in bulk).

A pendulum rod is no exception, and the consequence is that an increase in temperature causes a clock to lose time on account of the increased length of the pendulum rod. Of the common metals, steel expands the least, and is therefore the most suitable for a rod. Brass would nearly double the error.

In the case of a pendulum there are three factors to take into consideration. (1) The expansion of the steel suspension spring. (2) The expansion of the rod. (3) The expansion of the bob.

The first two of these tend to lengthen the pendulum. The last tends to counteract this lengthening by expanding upwards, as the bob rests upon the rating nut.

So if a pendulum were made with a steel rod expanding downward as 14, and a lead bob expanding upwards as 35, the upward expansion of the bob would to some extent compensate for the downward expansion of the rod, the amount depending on the length or height of the bob. But it is impossible to make a lead bob high enough to perfectly compensate the pendulum in this way. To do so, the bob would have to be nearly the length of the pendulum rod. For though 14 in. of lead will expand as much as 35 in. of steel, it must be remembered that a total lengthening of the bob of $\frac{1}{10}$ in. only raises its *centre* $\frac{1}{20}$ in., and therefore to compensate a 35 in. steel rod the lead bob would have to be 28 in. high (approximately— this being only a rough illustration, the "centre of oscillation" not coinciding with the centre of length).

Still zinc and steel compensation pendulums are made on this plan of counteracting the downward expansion of the rod by the upward expansion of a tube of zinc. Fig. 68 shows the usual arrangement. The centre pendulum rod A is of steel. At its lower end is a collar or rating nut E, upon which rests a zinc tube B. Upon the top of B is another collar from which hangs an outer steel tube C. To the bottom of C, the lead bob D is fixed. A moment's consideration will show that the expansion downwards of the centre rod A, and the outer tube C, tend to lower the bob and lengthen the pendulum, while the expansion of the zinc tube B is upwards and tends to raise the bob. The expansion of the lead bob is also upwards. So if the combined lengths of the steel rod and tube are as .34, to the length of the zinc tube and lead bob (to centre) 14, the pendulum will be compensated and the bob will remain at the same height.

F<small>IG.</small> 69.—Mercurial Compensation Pendulum.

F<small>IG.</small> 68.—Zinc and Steel Compensation Pendulum.

On the same principle a pendulum may be constructed of steel and brass, but it is not often seen in modern clocks, being springy and rather erratic in action.

A form of compensation pendulum often seen is the mercurial, shown in Fig. 69. A steel rod A has at its lower end a kind of steel stirrup B. In the stirrup stands a glass jar of mercury C. It compensates in this way : the rod A and stirrup B expand downwards and lengthen the pendulum. The mercury in the glass jar expands in volume and rises in the jar to a higher level, thus raising the centre of gravity of the bob. By proportioning the height and volume of mercury correctly, it will just compensate the pendulum as a whole. In practice a glass jar of 2 in. internal diameter and containing 6½ in. of mercury suffices.

The fault of all mercurial pendulums is that the rod being thin takes the temperature of the air more quickly than the jar of mercury (generally about 12 lbs. of mercury), and thus a rise in temperature for an hour or so may cause the clock to lose through the rod expanding and the mercury not becoming warmed quickly enough. This pendulum also has other disadvantages. It is expensive to make on account of the work in the stirrup and the cost of mercury, and mercury is awkward stuff to carry about or to have in a workshop, rendering such a pendulum difficult to handle.

The zinc and steel pendulum also has disadvantages, mainly constructional; the number of points of contact between the tubes causes a little uncertainty in action, and zinc is a soft and unsatisfactory metal to deal with, especially under heavy pressure of a large lead bob.

This brings us to " Invar," which is a steel and nickel alloy invented about 1900, that has the special merit of hardly expanding at all in heat. A plain pendulum with an invar rod will have less error than many compensation pendulums of the zinc and steel or mercurial forms. It is quite simple and cheap to compensate the small amount of the expansion of the invar rod by the upward expansion of a cast-iron bob resting on the rating nut. A pendulum of this type, as shown in Fig. 70, is the most perfect compensation pendulum that can be made and also one of the most permanent, being portable, cheap, and not liable to injury. The invar rod A is provided with a rating nut C,

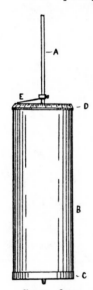

FIG. 70.—Invar Pendulum.

upon which rests the cast-iron cylindrical bob B. For a seconds pendulum, a bob 8 in. high fully compensates, and when of from 2 to $2\frac{3}{4}$ in. diameter gives a suitable weight for the pendulum of a good clock (about 7 to 12 lbs.).

A sort of "semi-compensation" pendulum is the wood rod and lead bob. Wood expands very little in heat, and a wood rod with a lead bob resting on the rating nut compensates fairly well, but has other errors caused by atmospheric influence upon the wood. Though a great improvement upon an uncompensated pendulum, this form is not equal to a properly made zinc and steel or mercurial pendulum and far behind invar. It is found in Vienna Regulators and in some English Regulators when the makers desired to save the expense of a mercurial pendulum.

CHAPTER VII.

PORTABLE CLOCKS.

French Carriage Timepieces.—Fig. 71 shows the movement of a French carriage timepiece with watch escapement and balance. A pendulum being out of the question in a clock designed for carrying about, resort is had to a balance and

Fig. 71.—French Carriage Clock Movement.

hairspring similar to that in a watch. So it may be said that up to the fourth wheel the movement illustrated in Fig. 71 is a clock, and beyond that it is a watch. The cleaning and repair of these clocks is really work for a watchmaker, but a few directions will be here given, by following which a clockmaker may be enabled to clean one.

These clocks have going barrels, wheels, and pinions exactly like the movement in Fig. 57 (p. 54), and Fig. 71 shows the arrangement of the train very clearly, the barrel being at the bottom, and the "platform" containing the escapement, across the top. They are fitted into cases made of brass, nicely gilt, and have glass sides and top enabling the movements to be seen. To take one apart the bottom of the case must first be removed. Stand the clock upside down upon a leather on the board, to prevent scratching of the gilding, and remove the small centre screw from the case bottom. This holds a thin sheet of brass which acts as a sort of false bottom; when this has been removed, four large screws will be seen at the corners going through into the pillars of the case. Remove these and lift out the movement bodily, still attached to the base of the case, as in Fig. 71.

The hands and dial may be next taken off and the main-spring let down with a key, holding the click back as usual. Or the pivots may be oiled, a wire put between the arms of the fourth wheel to prevent it running, the platform screws taken out and the platform removed. The clock may then be allowed to run down. But letting the mainspring down is the best method. Then take out the two large screws in the base and detach the frame. Take it all apart, removing everything from the plates, and clean and put together again as the French timepiece, Fig. 57.

The only portion of the train requiring special mention is the fourth wheel itself. This is termed a "crown wheel," the name being suggested by its form. The teeth are hori-zontal, that is, at right angles to the wheel, so that they may drive the upright escape pinion. An engineer no doubt would effect this with bevel wheels, but in clockwork where pressures are light and a wheel drives a pinion, the crown wheel system is found to be much superior in that the depth can be adjusted easily, and inaccuracies and difficulties in cutting are avoided. There is also not so much tendency for one to be forced away from the other by the driving pressure.

The depth is regulated by a hard steel screw with a polished point, inserted in the back plate. This takes the end thrust of the back fourth wheel pivot, and by its means the fourth wheel is kept up to its work in the escape pinion. Some adjustment is also possible by loosening the four platform screws and pushing the platform bodily over towards or away from the crown wheel. In oiling the clock there is a tendency to overlook this back crown wheel pivot, so it is best to place a little oil in the pivot hole before putting the frame together.

After polishing the plates, handle them with tissue paper or a leather to prevent finger marks and corrosion.

The platform and escapement may now be taken in hand. Remove the balance cock with balance attached. Then unscrew the escape cock that holds the escape wheel, and if a lever, the pallets also. Clean the platform and the cocks like the rest of the clock, if polished. If gilt, as they often are, rinse in petrol and brush them dry with the watch brush and dry chalk only. Sometimes the gilding is very tarnished ; when this is the case dip in petrol again, and wet the brush end in petrol also and brush the tarnish off. Dip the escape wheel and pallets in petrol and brush dry in the same way, holding them in tissue paper during the brushing process. Peg out the jewel holes, removing any endstones, which must be cleaned by rubbing face down on leather. When all is clean and pegged out, screw on the endstones again.

The escape wheel and pallets may now be placed in position again and the cock screwed on. If there are endstones, place oil in the jewel holes before putting the parts in place, and note that for all parts of the escapement, only *watch oil* should be used.

Oil the pivots, top and bottom, and screw the platform on the clock frame, adjusting the crown wheel depth very carefully.

The balance and hairspring comes next and needs especial care. Turn the buckle that bridges the curb pins, if it has one, then detach the hairspring stud from the balance cock. This may be screwed on or pushed friction-tight into a small hole in the cock. Do very little to the balance and spring, except to clean its pivots with pith. If a cylinder, peg the cylinder-hollow out clean. If a lever, brush the roller clean and bright, taking care not to touch the hairspring with the brush. Unscrew the index from the cock and remove the jewel endstone. Clean the cock with petrol and watch brush, peg out the jewel hole, clean the endstone and screw all together again. After placing a small drop of oil in the jewel hole, refix the hairspring stud. Then oil the top pivots of escape wheel and pallets and the foot hole of the balance pivot, place a little on the escape where teeth and put the balance in position again, screwing the cock down gently to see that it does not nip the balance. The balance must have just a little endshake or lift, sufficient to be seen with an eyeglass, but no more. It can be regulated by placing a slip of tissue paper under the forward end of the cock to increase it, or at the back end to decrease it. The outer coil of the hairspring should pass between the index curb pins and play evenly between them as the balance vibrates. It should not bind hard against one and not touch the other. When the balance is at rest the spring should just lie freely between the pins without touching either. See also that the hairspring lies flat and does not touch the arms of the balance or the stud or curb pins.

Really it requires a treatise on watchwork to fully deal with carriage clock escapements, and the reader is referred to "Watch Repairing, Cleaning and Adjusting," which gives directions for the adjustment and repair of these parts.

There is just one simple breakage that may be here referred to, and that is a cracked endstone. These are easily replaced, and must on no account be passed, as they damage the pivots and cause stoppage.

The clock being wound and going all right, put on the dial and hands again. Remove the side glasses from the case and clean them all up well and replace them in their grooves. Attach the movement to the base by its two screws and standing the case upside down on a leather, lower the movement into it and insert the corner screws, etc. Finally rub the gilding with a clean leather to remove finger marks.

These are very nice little clocks, though made in various qualities. The best are beautifully finished and keep good time. Cheap ones have a tendency to be a little rough where the workmanship cannot be seen, and often do not keep even time throughout the week. This is due to their rougher finish requiring a stronger mainspring and to the absence of "stop-work."

Stop-work is to arrest the winding before the last turn of the spring is reached, thus avoiding excessive strain. It also stops the clock before the last weak turns of the spring are reached. With a well-made clock and a thin mainspring making about eight turns, the stop-work enables the middle four or five alone to be used which gives much more even power during the week's run, than if the whole eight were used. Fig. 72 shows the arrangement. A is the "stop-finger" upon the barrel arbor. B is the star-wheel pivoted upon the barrel cover. Each time the stop-finger comes round, it moves the star-wheel one notch until the solid part C is reached. The available number of turns is then limited by the number of notches in the star-wheel, which is sometimes four and sometimes five according to the design of the clock.

FIG. 72.—Stop-work.

Suppose the spring gives seven turns and the stop-work is designed to use five. Then before putting on the stop-finger the star-wheel is placed as shown in Fig. 72 and the spring wound a turn. Then the stop-finger is put on and the star-wheel holds it permanently "set up" one turn. Similarly the stop-work will arrest the winding one turn from the top. A convenient way to effect this is to screw the barrel arbor in the bench vice and turn the barrel by the hand one turn round, holding it while the stop-finger is pressed home. A barrel with stop-work should never be allowed to run down quickly or the shock will damage the star-wheel or perhaps force it off its screw.

American-pattern Lever Clocks.—Fig. 73 shows the movement of a lever drum alarm clock, which is typical of its kind. These clocks as a rule are in cases about 4 in. diameter, but both larger and smaller sizes are found. They are also made to go thirty hours and eight days, and may have alarms or not.

In construction and design they resemble the pendulum clocks as far as the escapement, and may be cleaned and repaired in the same manner.

To take an ordinary 4-in. drum clock out of its case, first unscrew the feet and the screw at the top, or the alarm bell if it has one. The stem upon which the bell is fixed will also unscrew and can be gripped by cutting nippers, as may the feet if they are very tight. Having removed the set-hand button by pulling it straight off, the alarm button by unscrewing, and the winding handles by unscrewing in the reverse direction to that of winding, the case back may be prized off, and the movement drawn out. Sometimes the movements fit very tightly and the alarm hammer gives a little trouble, but they will come out with a little humouring and perhaps bending the hammer stem a trifle.

When out of its case, take off the hands. The minute hand may be removed by inserting cutting nippers underneath the

FIG. 73.—American Drum Alarm Movement.

centre boss and forcing and pulling it off—it does not do to be too gentle in taking these clocks apart. The hour hand, alarm hand, and seconds hand may be pulled off by the same means, but as a rule come off much more easily. Unbend the pins that hold the dial and take it off. The dial requires more careful handling and must not be bent or scratched, or marked by oil, or it will quite spoil the clock's appearance. Then detach the cast-iron frame to which the movement is screwed and it will appear as in Fig 73.

The first part to take out is the balance. Unpin the outer end of the hairspring and draw the outside coil through. Then loosen one of the centre screws in which the balance pivots work, and lift out the balance complete.   It is a little awkward to let down the mainspring of one of these clocks, so it is best to let the train run.   Place a clamp (Fig. 54, p. 49) upon the spring, after winding it up tight, then loosen the pillar nut or pin next to the pallets, and holding the escape wheel from running by placing a finger tip against its teeth, lift out the pallets and screw the plate down again.   The train may now be allowed to run until the spring fills the clamp.   If the pallets are held by a small cock, it is easier to take them out ; if not, loosening one pillar nut and carefully springing the plate up allows them to be removed.

To thoroughly clean the clock, take it all apart and place the parts in petrol as usual.   Many do not consider these

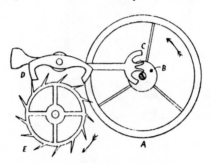

FIG. 74.—Drum Clock Lever Escapement.

clocks worth taking apart, and after removing the pallets and motion wheels, wind the spring up and place the whole movement in petrol to run down and wash itself and its pivots clean, dabbing it with a brush as it does so to remove dirt, etc. When run down throw off all loose petrol by swinging the movement and dry it with a duster.   In a few minutes the petrol will have completely dried off and the clock completed. By either process the pallets and balance get a good cleaning, and that is the essential part.   Before the pallets can be replaced properly the principle of the escapement must be mastered.   Fig. 74 shows the arrangement.   A = the balance, B the impulse pin, C the lever horns, D the pallets, and E the escape wheel.

The pallets are of the dead-beat variety, so the depth must

be adjusted till the teeth points just "lock," as shown in Fig. 74. (See also Fig. 62, p. 61.) When there is a small movable cock to hold the top pallet pivot, adjusting the depth is easy. When there is not, the top pivot generally is held by a narrow fixed tongue of metal, a part of the stamped-out plate, which can be bent a little either way by a strong pair of pliers. The escape wheel teeth should only just "lock" and no more.

The centre screws of the balance must be pegged out clean and the balance pivots examined with an eyeglass. They should be absolutely sharp points, with no rounding off or one-sided wear visible. If not, hold the balance axis in a pinvice and carefully point it up on an oilstone until the centres are true and sharp. Then put the balance in, seeing that the impulse pin goes in the centre notch of the lever fork, adjust its centre screws so that it is quite free and has hardly any perceptible shake. It *must* have a little, but should not be allowed to rock about. Re-pin the hairspring in its stud in such a position that when the balance at is rest, the impulse pin is in the centre of the lever fork, and lever, pin, and balance are all in straight line. The clock will then be in beat.

Oil the balance centres, the impulse pin, the pallets, all pivots and the coils of the mainspring, wind up the clock, and it should go perfectly.

If not, look for small faults and begin at the escapement. Make quite sure that the balance pivots are really sharp points and that the hairspring is flat and free from the balance arms or the plate. Then see if the lever is free. Hold the balance with the finger tip, about half a turn round from its position of rest, and see if the inner lever horn binds against the balance axis. It should not do so, but should have a little shake. Then turn the balance round the other way and try the same thing on the side of the lever. If the lever horn on one side binds and that on the other has a great deal of shake, the lever must be bent a little to one side, or if it is provided with banking pins, one of these should be bent to give freedom, as the clock cannot go with the lever horn dragging against the balance axis all the time and acting as a brake.

The motion wheels and dial may be replaced and the hands put on. If the clock has an alarm, the hands must be put on in the correct position or the alarm will not be right. The simplest way is to wind up the alarm a little and turn the set-hand arbor round until the alarm goes off. Then put on the alarm hand at, say, 6 o'clock and put the hour and minute hands to correspond so that they show 6 o'clock also when the alarm goes off. They will then be correct for any other time. If the alarm hand will not stay where it is set, but travels as the clock goes, it shows the alarm set-arbor moves too easily. It is generally held by spring washers and a pin, and having been unpinned, the washers may be curved a little more to make them hold firmly. It is very much like the set-hand arbor, and when cleaning a clock, both arbors should be tested for stiffness before putting the clock together. The alarm wheel under the dial has a sort of cam upon it, on which rests a pin fixed in the alarm arbor. A notch in the cam allows the alarm wheel to rise for the alarm to go off. This cam should be oiled to prevent harsh friction.

American Carriage Clocks.—Movements of the same character and of much the same form are found in metal cases after the style of French carriage clocks, fitted with glass sides and handle at the top, etc., and similar movements, but smaller, are found in fancy china and wood cases of all descriptions, making cheap little portable clocks that may be carried from room to room. All are on the same plan and may be treated in the same manner.

Also many of the so-called watches that are sold for a few shillings are simply diminutive drum clocks of the same pattern and workmanship.

There is a variation of the escapement found in many, and instead of having solid steel pallets as in Fig. 74, they have pin-pallets and inclined impulse planes on the escape teeth. The principle is exactly the same. The teeth must just lock by dropping upon the pallet pins just below the corners, the "locking faces" being on the teeth instead of on the pallets. One advantage of these pallets is that they can be replaced when broken or worn. A small punch made from a needle with the point filed off, and held in a pinvice, will drive out a broken or worn pin and a new one may be made from another needle, which being ready hardened, tempered, and polished leaves little else to be done.

CHAPTER VIII.

ENGLISH STRIKING CLOCKS.

One at the Hour.—The simplest kind of striking work is that found in some English skeleton clocks which strike one blow at each hour, just as a reminder. Fig. 75 shows the mechanism. The cam or snail, A, is on the cannon pinion and revolves in the direction of the arrow. B is the lower arm of the hammer and rests upon A. The position shown in Fig. 75 is about five minutes to the hour; when the snail A travels a little further, the arm B will fall and the hammer head C will strike one blow upon the bell D. The movement of the hammer is assisted by a light spring.

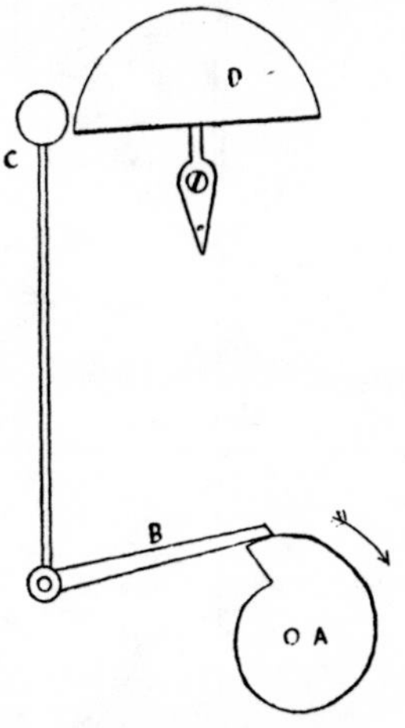

Grandfather Clocks. — To strike the correct number at each hour requires special mechanism and a separate train of wheels. There are several systems on which such mechanism is arranged. A common form, and one of the best,

Fig. 75.—One-at-the-hour Striking Mechanism.

is what is known as "rack-striking work." The ordinary English eight-day, long-cased clock is a familiar example. All of these, except a few very old ones, have rack striking work.

Fig. 76 shows the striking mechanism of a "grandfather" clock. The parts shown are arranged upon the front plate and

G

underneath the dial. The wheels between the plates are indicated by plain circles just to show their position.

The action of this mechanism will be first described. A is the main wheel or barrel similar to that of the going train. This drives the " pin wheel " B, so called from the eight pins it has around its rim. These pins lift the hammer tail L, and when a pin passes, it lets L fall and the hammer head M gives

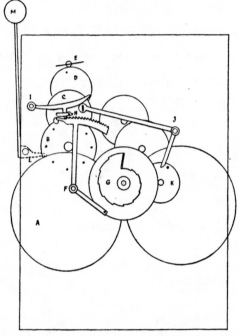

FIG. 76.—English Rack-striking Mechanism.

the bell a blow. This is actually the striking work, and it is operated by a separate weight like the going side of the clock. As each pin of the pin-wheel passes the hammer tail, one blow is struck. The rest of the mechanism is to regulate the number of blows struck at each hour and the speed of striking.

The next wheel in order in the striking train is the "pallet-wheel" C, which carries on a prolongation of its front pivot the "gathering-pallet" H. The pallet-wheel makes eight revolutions to one of the pin wheel, or one for each pin. So for each blow struck the pallet-wheel revolves once and the gathering pallet on its front pivot also revolves once. The "snail" G is fixed upon the hour wheel and revolves once in twelve hours. It is a disc of brass divided into twelve sections, each section being a step lower than the one before. In conjunction with the rack F, this regulates the number of blows struck at each hour.

The next wheel in the train D is the "warning wheel." This has one single pin in its rim, the warning pin. The last of the striking train is the "fly" E. This is simply a fan, and its sole purpose is to regulate the speed of running, which it accomplishes by the resistance the air offers to its rapid revolution. Its regulating power depends upon its area and not on its weight. The larger and lighter a fly is made, the more effective it is.

Other parts of the mechanism are the "rack hook" I, which is detent designed to allow the rack to be gathered forwards tooth by tooth by the gathering pallet H, but prevents the rack from falling back, and the "warning lever" J, whose business it is to let off the striking just at the hour.

The whole mechanism acts in this way: as the hands approach the hour, a pin in the minute wheel K lifts the lower tail-end of the warning lever J. In lifting, the long upper arm of J lifts the rack-hook I, which lets the rack F fall until its lower end, the "rack-tail," falls against one of the steps of the snail G. It is obvious that the distance the rack falls is regulated by the depth of the step of the snail against which its tail rests. When the tail rests upon the deepest step, the rack falls a distance equal to twelve teeth. When upon the highest step, the distance fallen is equal to one tooth.

The falling of the rack releases the striking train, by the pin in its upper extremity slipping away from underneath the wing of the gathering pallet. The striking train runs a little way only, just until the pin in the warning wheel D comes round and is stopped by the stop block in the end of the warning lever J, which goes through a slot cut in the frame plate, and when J rises, intercepts the warning pin. This short "run" of the wheels is termed the "warning," and generally takes place about five minutes before the hour. Its purpose is to liberate the striking train and prepare for letting it off exactly at the hour. It does this as follows: It should be remembered that after the "warning," the rack rests upon the snail; the gathering pallet is free to turn round and gather up the rack teeth; the striking train is merely held from running by the fact that the stop block on the upper arm of the warning lever stops the warning pin by blocking its path, and the warning lever is held by the pin in the minute wheel advancing. This pin continues to advance, and just when the minute hand points to the hour, the pin passes the point of the lower arm of the warning lever, letting it fall. The falling of the warning lever accomplishes two things, it lets the rack hook fall into position to hold the rack and liberates the warning pin, allowing the striking train to run and the clock to strike. This it does. At each revolution of the pallet wheel C and gathering pallet H, one blow is struck upon the bell and one rack tooth is gathered up. The striking continues until the rack teeth are all gathered up and the wing of the gathering pallet again comes to rest upon the pin in the end of the upper part of the rack.

This is the whole operation of striking the hour, and it should be studied most carefully. Once mastered, no difficulties will be found in cleaning or repairing these clocks. The best way to familiarize one with every movement, is to take a clock erected and going, remove the hands and dial as it stands, pin on the hands again properly, and turn them slowly towards the hour. Watch what happens at the "run" or "warning." Watch the warning lever lift, the rack fall, the train run, and the warning pin stop against the stop block of the warning lever. Then continue the forward motion of the hands until the warning lever falls and liberates the train. After this point, things move too fast to watch effectively, so place the finger tip upon the fly and only let it turn half a turn at a time. Then the exact way the gathering pallet and rack act may be seen, until the gathering pallet finally comes to rest upon the rack pin.

A grandfather clock is taken to pieces and cleaned exactly as an eight-day English dial clock. It is a little more difficult to put together because of the larger number of wheels, etc., but if the top plate is put on and the long pivots got in their holes, the winding squares, centre arbor, seconds pivot and pallet wheel pivot, the pins may be put in the lower pillars to hold that end of the plate down, and the rest got in one by one until the plate drops on flat.

Then put in one upper pillar pin, put on the rack, rack hook, and gathering pallet, not pinning any of them. Cause the striking train to run by pulling the line of the striking barrel, and continue until the rack is gathered up and the gathering pallet rests on the rack pin. Then look at the hammer. It should have just struck a blow and its tail should be quite free of the lifting pins in the pin-wheel. If it is half lifted for the next blow, take off the gathering pallet and replace it on another square and try again. If none of the four positions is correct, one being just too soon to let the hammer fall, and the next not stopping the train until the hammer begins to be lifted for the next blow, it shows that the pin-wheel must be shifted one tooth round in the pallet wheel pinion. To do this, take off rack, rack-hook, and gathering pallet, remove the upper pillar pin and raise the plate sufficiently to get the top pivot of the pin-wheel out of its hole. Then turn it one tooth and push back, taking care that the pallet wheel has not moved. This has the effect of advancing the pins of the pin wheel without advancing the gathering pallet.

Put on the rack, etc., and try again, and a position should be found in which, when the gathering pallet stops, the hammer has just struck and is quite free. If a clock is passed with the hammer " on the lift," the striking will fail to go off, as the train just at starting must have no resistance offered to it. When once on the run it will overcome a great deal, but it will refuse to start unless quite free.

Having got this part right, see where the warning pin is. It should be half a turn at least from the stop-block of the warning lever so as to have half a turn of " run " at the warning. If not, raise the plate again and shift it round. This is simpler than moving the pin wheel and is easily done. Get out the warning wheel bottom pivot and bring its pinion clear of the pallet-wheel teeth, turn it round, and replace.

There are the two points to look to in a rack-striking train. First, the hammer must have just struck a blow when the train stops and must be quite free from the lifting pins.

Second, the warning pin must have not less than half a turn of "run" to the warning lever stop block. Then all will be well.

Several little faults sometimes occur in the striking work, chiefly in connection with the snail, rack, and gathering pallet action. Sometimes the rack tail gets bent, or moves upon the rack and causes trouble. This may make the clock strike one more than it should or one less, and the remedy is to bend it back or move it, until when on the one o'clock step of the snail it strikes one, and when on the deepest step, twelve. Then, if loose, fix it with solder or rivet it on tight. If the clock only strikes one too few or one too many at one or two hours, it is probable that it nearly does so at all the rest, and a slight movement of the rack tail will correct them all.

In putting on the hour wheel and snail, take care that at the hour, the snail is in position so that the pin in the rack tail falls upon the *centres* of the steps.

When the clock stops between twelve and one o'clock, and the hands seem fixed, it will be found that the rack tail has jambed tight against the long step of the snail. The cause of this is a failure to strike. The rack has fallen and the clock not having struck, the rack tail has remained until the long step coming round, it has stopped the clock. This is a frequent cause of damage to the rack tail. It indicates a fault somewhere in the striking work, such as bent pivots, a hammer left "on the rise," or a catching of the gathering pallet on the points of the rack teeth.

The latter is a common fault. If the rack falls the least trifle too far, the gathering pallet point may catch on the point of the tooth. The remedy is to bend the rack tail a trifle. Such faults can be located by placing the finger tip on the fly and only allowing it to turn a little at a time as before described while the action is carefully watched.

The pivots of the striking train should be kept smooth and straight, the holes well bushed, or else a disagreeable rattling sound will be heard during striking. The fly should be held friction-tight upon its pinion by a spring. If this is too loose, it allows the pinion to run round without the fly and the striking is too rapid. If too tight, there is a sudden stop and a rebound when the train comes to rest, instead of a smooth, steady stop. To tighten a fly spring, the fly must be taken off its pinion, it cannot be done when the clock is together, so should always be tested when cleaning.

To make the clock strike exactly at the hour, the minute wheel may be shifted in the cannon pinion until correct, and to bring the falling of the rack tail pin to the centre of the steps of the snail, the hour wheel may be shifted in the leaves of the minute pinion a tooth or two.

The rack spring should be fairly strong so as to make the rack fall sharply; if weak, sometimes the rack rebounds from the snail and the rack hook catches it on the rebound, making the clock strike one or two short. This is especially liable to occur at the longer hours, nine to twelve o'clock, when the rack falls a considerable distance, and happens when the hands are set forward to set the clock to time.

In oiling these clocks, oil the rack pin where the gathering pallet wing rests, the minute wheel lifting pin, the warning pin, and the pin wheel lifting pins. Also the point of contact of the rack spring and the hammer spring.

A dry or rusty pin in the rack often will prevent the rack falling by causing harsh friction at the gathering pallet wing. A dry or rusty warning pin will prevent the warning lever falling freely.

Finally, before setting one of these clocks up and leaving it, strike it round twelve hours to test the mechanism thoroughly, and when putting the hands forward quickly, pause at the warning to give the rack time to settle, or the clock will strike a short number.

In setting up a grandfather clock, see that the pendulum bob is quite free from the case back. This may be attained by tilting the movement forward a little or shifting its seat board forward. Failing these methods, place something behind the back of the clock case to tilt it forward a little. The case should stand quite firm and if possible be fixed to the wall by a screw to steady it.

The simplest and most accurate way to set the clock in beat is to bring the pendulum to rest and make a pencil mark on the case back behind the end of the rod. Then draw it gently to either side and make two other marks where it "beats." These should be equi-distant from the central one, and if not, the crutch must be bent until they are so.

A new line may be generally put on as the clock stands,

by turning the barrel round until the line hole is outermost, and inserting the new gut and drawing it through with a hook. A knot is tied and the end burned to prevent it untying, then drawn and pushed back in its hole.

Most grandfather clocks go eight days, but a few old ones only go thirty hours. These are arranged upon a different plan and as a rule have one weight for both going and striking trains. This weight hangs on an endless chain or rope and is kept taut by a small secondary weight to prevent slipping. They are wound by drawing up the weight by means of the line. A frayed rope should be replaced by a new one, which must be spliced or halved and sewn together. Special soft ropes for them can be bought at clock material shops. The wheel spikes must be kept sharp.

The striking mechanism of these clocks is often on the " locking plate " system, which will be described later on.

This is not a history of clocks and clockwork, or much that is of interest might be said about these clocks, their age and their makers; how the pattern and form of the dial reflects their age and the advancement of the art of timekeeping; how they have developed from the old " one hand " clock with hours divided into quarters, etc.

Bracket Clocks.—The same may be said of English bracket clocks. These have the same pattern of movement and striking work, the older ones " locking plate," and the newer ones racks, and they will need no special directions to clean and repair.

Locking-plate Striking Work.—Fig. 77 shows the back plate view of a locking-plate striking mechanism. A is the pin wheel which lifts the hammer. B is the next or " cam " wheel, having a brass disc or cam upon its axis, in which is an opening for the arm G to fall into, and stop the striking train. C is the warning wheel, D the fly. F is an arm attached to the same axis as G and resting upon the rim of the locking-plate E. H, I, and J are three arms attached to one axis and acting as one. J is outside the front plate and lifted by the pin in the minute wheel as the hour is approached and let fall again at the hour. I is the warning arm and a block on its end catches the pin in the warning wheel C. H extends downwards and is bent at a right angle to pass under G and lift it.

The locking-plate is the portion that determines the number of blows struck at each hour. It is a disc of brass, mounted upon a wheel which is turned once in twelve hours by a pinion on the outer end of the pin-wheel axis. The disc has twelve deep slots at distances equal to 1, 2, 3, etc., up to 12, the first

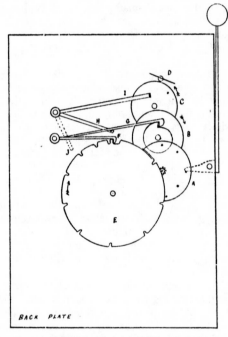

FIG. 77.—English Locking-plate Striking Mechanism.

and second slots being so close together as to run into one slot of double width.

The action is as follows: As the hour is approached, the advancing pin in the minute wheel lifts J, H and I also rising as they form one piece. I comes in the path of the warning wheel pin. H raises G and F. The effect of this is that the hook G is raised clear of the slot in the cam on B and liberates the striking train, which runs until I catches the warning wheel pin. The minute wheel advances until J drops at the hour, liberating the warning wheel and allowing the clock to strike, G resting on the edge of the cam as it revolves, and F resting on the rim of the locking-plate. Each time the slot in the cam comes round (once for each blow of the hammer) G would be free to fall and stop the train if F were not supported by the rim of the locking-plate. But F is supported by the locking-plate until a slot comes in position under it. So the clock strikes until a locking-plate slot comes under F, then G and F fall and G catches the cam-wheel B and stops the train.

There are many variations of this striking work, but the underlying principle is the same, and with the clock before one, it is easy to follow the action. To study it carefully, place the finger tip on the fly and let it only turn slowly while the action of the various parts is watched. If allowed to run in the ordinary way, they move too fast to follow with the eye.

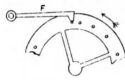

FIG. 78.—Pin Locking-plate.

Sometimes a locking-plate is made by drilling twelve holes at distances 1, 2, 3, etc., and inserting steel pins. Then the action is reversed, as in Fig. 78, the arm F being raised by each pin to stop the striking, and falling between them while the clock strikes. When this kind of locking-plate is used, the action of the other parts is modified to suit it.

As the pin wheel (A, Fig. 77) has eight lifting pins, if a pinion of eight leaves is fixed to it its back pivot outside the plate, one leaf of the pinion will represent one pin, or one blow of the hammer. So for the locking-plate to make one revolution in twelve hours, it must be mounted upon a wheel of seventy-eight teeth driven by the pinion of eight, seventy-eight being the sum of the numbers 1 to 12 inclusive, or the number of blows of the hammer in twelve hours. Mounted thus, the locking-plate may be looked upon as divided into seventy-eight spaces, with slots at the 1st, 2nd, 4th, 7th, 11th, 16th, 22nd, 29th, 37th, 46th, 56th, and 67th divisions. The 1st and 2nd will merge together forming one wide slot.

In putting one of these clocks together, the same two points must be considered as in rack clocks. First, the hammer must have just struck one blow when the train stops and must not be partially lifted for the next blow. The hammer tail should not touch the lifting pins. Second, the warning pin must be at least half a turn from the stop block of the warning lever.

The striking train will stop when the arm G falls into the slot in the cam B, so the pinion of B must be placed in such a position in the teeth of the pin-wheel, that when the hammer tail escapes from a pin, the cam is in position to stop the train, and if not correct at the first time of putting together, the plate must be raised and the pinion shifted as regards the teeth of the pin wheel. To get it quite correct, it may have to be shifted several times, but it must not be passed until correct.

When this is made right, turn to the warning wheel and see that its pin is half a turn from the warning lever stop block when the train is locked. If not, shift it by raising the plate until its top pivot can be got out of the hole and give it the necessary turn round before slipping it in again.

Sometimes after cleaning one of these clocks, it has a tendency to strike one too many, or to strike two hours without a break, the arm F not seeming to fall clean into the slots of the locking-plate. When this happens, try the locking-plate in a different position on its arbor, also try the pinion that drives it in a different position until one is found in which the striking is correct. There is always a little want of perfect truth in the best handwork, and these clocks are all made by hand, the parts being often only true when on their arbors in the position in which their makers fitted them.

To many not brought up to the business of clock making, striking work is a puzzle, but it will be quite simple if the two essentials are borne in mind, viz. the hammer must not be " on the lift," and there must be half a turn of " run," and by following the instructions in this chapter, this may be attained.

CHAPTER IX.

FRENCH AND AMERICAN STRIKING CLOCKS.

Ordinary French Rack Striking.—French movements like that shown in Fig. 57 are made both with rack striking and locking-plate. Fig. 79 shows the arrangement of a French rack striker. When compared with the English form, many

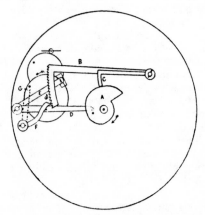

Fig. 79.—French Rack-striking Mechanism.

points of difference will be seen. The snail has no "steps," but is merely a continuous curve. The rack is placed horizontally and falls by its own weight only. Also the clock is arranged to strike one at the half-hour.

The mechanism acts as follows : As the hour approaches, a pin in the cannon pinion lifts the arm D, to which is also fixed the warning arm E. At the top end of E is the usual stop block passing through a slot in the plate and engaging with the pin in the warning wheel. A pin in the rack detent F also rests upon D, so that when D is raised, the detent F rises also and lets the rack B fall upon the snail A. The detent F is squared on to the front end of an arbor that goes through the frame plates and carries the arm G which has a stop block at its end, engaging with the locking pin in the pallet wheel as in Fig. 80, this being between the plates. Therefore, when the detent D is raised and the rack falls, G is raised also and liberates the pallet wheel. The train then runs until the warning wheel pin is stopped against the block on E. This is the warning. The pin in the cannon pinion continuing to advance, at last leaves the point of D and lets it fall. This effects two things. The warning pin is liberated, allowing the clock to strike, and the detent F falls against the rack teeth and supports

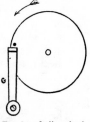

Fig. 80.—Pallet-wheel Locking Arrangement.

it as it is gathered up tooth by tooth by the gathering pallet. When the last tooth is gathered up, F falls under the end of the rack and G intercepts the pin in the pallet wheel as in Fig. 80, and the striking train comes to rest.

To strike one at the half-hour, there is another pin in the cannon pinion placed a little nearer to the centre than the one which lets off the hour striking. This lifts D not quite so high, but just high enough to raise F and G until the pallet wheel pin is liberated and the warning takes place, but not high enough to let the rack fall. At the half-hour D, E, F, and G fall, and the pallet wheel runs one revolution until it is again stopped, as in Fig. 80.

As in all striking work, there are two things to bear in mind when putting it together. The hammer must not be " on the rise " when G locks the pallet wheel as in Fig. 80. This is governed by the position of the pallet wheel pinion in the teeth of the pin wheel. One leaf of the pinion is filed to mark it, and in the rim of the pin wheel there is a dot against one tooth space. When standing the wheels up on the plate ready to put together, place the filed leaf of the pinion in the dotted space. The other point is that when the pallet wheel is locked, as in Fig. 80, the warning wheel pin must be half a turn from the warning lever stop block. There is a dot also in the rim of this wheel as a guide in pulling together, and when the pin wheel and pallet wheel are in position as before described, this dot in the warning wheel goes next to the fly pinion.

Even when carefully put together by these marks, sometimes it will not be quite correct, and in that case the plate must be raised and the wheel or pinion moved a tooth as required. Some clocks have a small cock or bar that is removable and enables the wheel to be moved without raising the plate. In no case must the plate be raised while there is any power on the train, so all the testing should be done before winding up. It is quite simple ; as soon as all the wheels are in, turn the pin wheel round once, let off the hammer tail and bring it to rest against the stop G. Hold it there while the freedom of the lifting pins is observed and also the position of the warning pin.

The greatest care should be taken not to bend the pivots when putting these clocks together.

The snail having no steps, it is important that it should be put on in the right position, or wrong numbers may be struck. A dot will be found on the cannon pinion against a tooth and another on the minute wheel against a space. Put these together, and while so, put on the hour wheel and snail with the dot in the minute pinion, then all should be right. But it must be carefully tested at 12 o'clock to see that the rack tail does not jamb against the long step of the snail, and also at 1 o'clock to see that the tail falls safely on the high part.

When one of these clocks is found stopped between 12 and 1 o'clock with the rack tail jambed against the long step of the snail, it shows that the tail did not have sufficient clearance when it fell at 12 o'clock, or else it indicates a fault in the striking mechanism that caused the clock to fail to strike. This failure need not have occurred at 12 o'clock, but may have been at any hour, the rack tail simply resting on the snail until the long step came round and stopped it.

There are many possible causes of failure to strike. Breaking of the striking mainspring, the striking side not being wound, a bent pivot or a bent tooth in one of the wheels, bad depths, a dry or rusty warning pin, a fly "out of poise" (heavy on one side, causing resistance to starting), a hammer

" on the rise," or the gathering pallet catching on the points of the rack teeth. The latter fault can be cured by moving the hour wheel a tooth in the minute pinion, thus letting the rack fall not quite so far, or a little further, as the case may be.

If the clock strikes the hour at both hour and half-hour, it shows that the half-hour pin in the cannon pinion lifts D too high and the remedy is to bend the pin inwards towards the centre of the cannon pinion. If it only strikes one at the hour, the hour lifting pin must be bent a little outwards so as to lift D higher and let F release the rack.

When the clock has finished striking, the proper position for the gathering pallet is pointing upwards.

Locking Plate.—Fig. 81 shows the mechanism on the

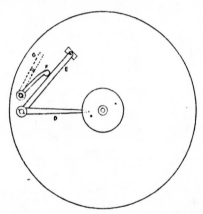

FIG. 81.—French Locking-plate Mechanism (Front Plate).

front plate and Fig. 82 that on the back plate of a French locking-plate striking mechanism.

As in the rack striker, D is the letting-off lever and E, in one piece with it, is the warning lever. F is an arm squared on to the locking lever G which is between the plates, and locks the pallet wheel as in the rack striker, as in Fig. 80. A projecting piece on G comes through a slot in the back plate and rests on the edge of the locking plate A, Fig. 82.

The locking plate is made on the same principle as the

English one before described, except that the spaces are each of double width, to enable one to be struck at each half-hour. In making such a locking plate, its rim is divided into ninety equal parts, two being cut out between each hour.

The mechanism acts thus : As the hour is approached, a lifting pin in the cannon pinion raises the lever D and warning arm E. F also rises, and with it the locking arm G, liberating the train which then " runs " until E catches the warning pin. At the hour, D falls, E falls also and liberates the warning pin and the clock strikes. As it strikes, the locking plate A slowly

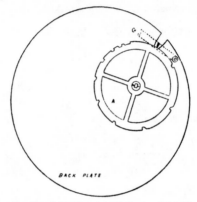

Fig. 82.—French Locking-plate Mechanism (Back Plate).

turns and the projecting stud on G, which rests upon its edge, cannot fall until it reaches a gap. It then falls and G catches the pin in the pallet wheel and stops the train.

At the half-hour, exactly the same thing occurs, but as the stud on G rests in a double gap on the edge of the locking plate, the pallet wheel is again stopped by G when it has made one revolution, and consequently the clock only strikes one.

This locking plate striking is very simple, but is subject to the same remarks as the rack clock. The hammer must not be " on the rise," and the warning pin must have half a turn of run. The same marks are there as guides in putting together, and the wheels are adjusted to one another in the same manner.

If from any cause, a locking-plate fails to strike, it simply misses an hour or a half-hour as the case may be and the striking becomes wrong. When it does so, look out for a fault as described in connection with rack clocks.

In French clocks, whether rack or locking-plate, if the clock strikes a little before or after the hour instead of just at the hour, the easiest way to correct it is to bodily twist the pipe of the cannon pinion so that the hand points to the hour when the lifting pin lets D fall. Then if the half-hour is not quite right, bend the half-hour pin a little forward or backward as required. But the hour is the most important. Sometimes the spacing of a locking plate seems not quite right, causing wrong numbers to be struck. In such a case try the plate in another position on its square until one is found at which all hours are correct.

Carriage Clock Rack Striking.—French carriage clocks generally have another form of rack striking, which is simpler and very certain in action. It also has the merit of enabling the clock to repeat the hour when a button is pressed correctly right up to the moment of striking. If a rack-striking clock of the ordinary form were provided with a button to raise the rack hook and let the rack fall for the purpose of "repeating" the striking of the last hour, it would not act after the clock had "warned," *i.e.* between about five minutes to the hour and half-hour and the time of striking. This is owing to the fact that E holds the warning wheel and does not release it until the time for striking. Also towards the hour, the clock would probably strike one too many owing to the snail having travelled and the next step having come round.

So a variation has to be made in repeating clocks to effect two things. First, there must be no "warning," so that the train is free to run up to the last minute. Second, the snail must not travel gradually, but must be made to jump a step just before the hour.

Fig. 83 shows the arrangement, and it is so good, simple, and certain in action that it constitutes a considerable improvement on ordinary rack striking in any form of clock.

First as regards the snail, it is mounted on an independent "star wheel" A, which has twelve long-pointed teeth and

H

revolves upon a stud fixed in the front plate. It is held in position by a spring "jumper" D, and is moved about one minute before the hour by a pin in the cannon pinion B. This pin engages one of the long teeth of A and gradually moves it forward until a tooth point of A passes the point of the jumper D. Then D completes the motion and drives A forward one division suddenly, and this is timed to occur just

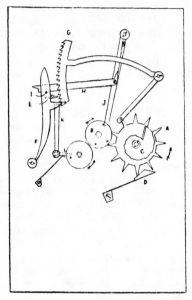

FIG. 83.—French Carriage-clock Rack-striking Mechanism.

before the hour-striking is let off. The exact time of jumping is determined by the position of the pin in the cannon pinion.

The rack G lies horizontally as in all French clocks, and the rack detent F is of the shape shown and has a pin I in its upper end. The letting-off of lever J works on a stud at its upper end and has a horizontal arm H jointed to it. At the end of H is a notch which engages the pin I in the rack detent.

It all acts as follows : As the hour approaches a pin in

the cannon pinion draws back J and H until the notch in H is drawn past the pin in F, upon which it rests. At the hour, the star wheel jumps, J is released and its spring causes H to advance, and the notch in its end catching the pin I pushes the rack detent F outwards and lets the rack fall. F is squared on to an arbor passing through the clock plates and carrying an arm which locks the pallet wheel (as G in Fig. 80), so that the same action that releases the rack, also releases the striking train, which immediately runs and the gathering pallet commences to gather up the rack. The wing E on the gathering pallet as it comes round raises H and lets F fall into the rack teeth. When the rack is all gathered up, F falls further in and the locking arm between the plates, attached to it, locks the pallet wheel as at G (Fig. 80).

To make the clock strike one at the half-hour, there is a lever LK. The arm L is depressed by a pin in the minute wheel, just at the half-hour, so that K comes under a pin in the lower end of the rack and only allows it to fall one tooth. The letting-off mechanism acts in exactly the same manner at hour and half-hour.

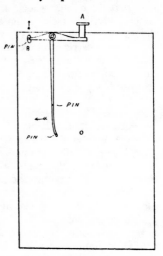

FIG. 84.—Repeating Lever.

To make the clock repeat there is a piece, Fig. 84, working on a stud at the top of the clock plate. A press button in the case depresses A. The long arm that goes downwards has two pins, one presses back the rack detent F and lets the rack fall, and the other holds back K so that the rack can fall quite clear even if K is in position to stop it. A pin at B going through the plate, catches the fly and holds it until the finger is taken off the press button, when the clock will repeat the last hour.

In carriage clocks, designed to act in any position and to perform even when packed upside down in a portmanteau, it is

essential that every part should act by a spring. Therefore
the rack is provided with a spring to make it fall, the rack
detent, the lever LK, and the repeating piece (Fig. 84) all
have springs, although when in an upright position most of
these parts would act quite correctly without them owing to
gravity.

In a repeating clock, it is necessary for the mainspring of
the striking train to be extra long, or to give an extra turn, as
if repeated many times, there would be a risk of the striking
train failing before the end of the week was reached.

When cleaning these clocks, as usual, see that the hammer
is not "on the rise" when the pallet wheel is locked.   As
there is no warning, it is simplified to that extent.   The rack,
the levers J, H, L, and K must all move quite freely and their
springs must be oiled a little at the points of contact.   All
lifting pins should be oiled and the face of the jumper D.   Take
care that the minute wheels are so put together that L is lifted
so as to stop the rack falling at the half-hour.   There will be
dots on the minute wheels for this purpose.

Being a carriage clock, all the plates, etc., must be nicely
polished, and it is necessary to remove every lever and spring
before polishing.   It is well to keep each spring with its
part and lay in one place on the workboard to avoid
confusion.

American Locking Plate Striking Work.—Fig. 85
shows a typical American striking movement.   The locking
plate C is generally mounted upon the front plate, though it is
sometimes upon the main wheel between the plates.   The
locking arm B dips into the slots in the locking plate and its
inside arm rests upon the cam G, as in an English locking-
plate clock.   A is the letting off lever and is between the
plates, its second arm E raises B, and its third arm F acts as
a warning lever.   Its lowest arm I is raised by the projecting
wire H on the centre arbor.

As usual, the hammer must not be on the rise when the
clock has just finished striking, and it may be regulated by
altering the relative positions of the cam wheel and pin wheel,
raising the plate and moving one a tooth or two as required.
The warning also must have half a turn or so of run and that
may require shifting also to adjust it.

There are no marks in these clocks for putting the striking train together correctly, so it must be put in by guesswork at first, tried before winding the spring, by forcing the pin-wheel round and causing a few blows to be struck, and altered until correct in these two main particulars.

The thin end of the dipping arm B must dip centrally into the slots in C when the cam wheel allows it, and B must be

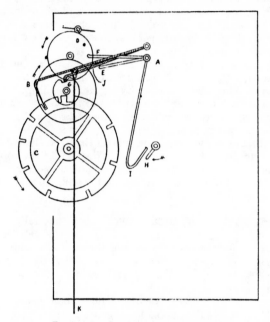

FIG. 85.—American Locking-plate Striking Work.

adjusted by bending at the corner or elbow. These arms being wire, sometimes get bent, so it is essential to understand their action to be able to bend them back again. It should be noted that the gap in the cam on the cam wheel is to let the upper arm of B fall so that the thin end of the lower arm may dip into the locking plate. These two arms should be adjusted so that when the thin end dips into a deep slot, the upper arm is well in the gap in the cam, and when the upper arm rests on the edge of the cam, the thin end of the lower arm cannot catch in the locking plate slots.   A little careful bending will effect this after one or two trials.   Then the lifting arm I must be bent until its lower end is released exactly at the hour.   When it falls its top arm should fall clear of the path of the warning pin, and when raised, the top arm should intercept the warning pin.   The middle arm E, when raised, should raise B out of the cam slot, and when dropped should let B fall right into the bottom of the cam slot.   A study of the English locking plate in the last chapter will help to its proper understanding.

There are many variations of this mechanism found in American and German striking clocks, but a study of the three forms of locking plate here described, the English, French, and American, will enable all variations to be easily mastered. The thin wire K is to enable the owner to strike the clock round to the correct hour.   Pushing it up raises B.

In all there is a locking plate, a lever that rests upon it and falls into its deep slots to stop the striking train.   In all, the letting-off lever, raised as the hour approaches, effects the "warning," and when it falls releases the warning pin and lets the clock strike.

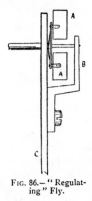

FIG. 86.—" Regulating " Fly.

Vienna Regulators generally have rack-striking work similar in pattern to English clocks, and the snail is mounted on a star wheel like that in Fig. 83, so these will need no special reference.

Regulating Fly.—Good clocks that strike upon large gongs are often supplied with large regulating flys, to ensure slow running and give a means of securing slow or fast striking as desired.   Fig. 86 shows such a fly.   The fly pinion is made of extra length and comes through the back plate C, its back pivot being held by a large " fly cock" B.   The pinion carries a cross arm and two square vanes, A, A, mounted upon studs that can be turned to any angle, like the fly of a musical box.

By this means greater air resistance is got, and by setting the vanes at an angle, the resistance may be modified as desired from almost nothing at all when set edgeways, to a maximum when set as in the figure. The cross arm carrying the vanes turns upon the pinion and is kept friction-tight by a flat spring underneath it, and a pin through the pinion body keeps it together.

Hammers, Bells, and Gongs.—Bell hammers should not fall direct upon the bells, but must be arrested by a fixed stop just before hitting the bell, the elasticity of the hammer stem and the stop allowing the bell to be struck. If the hammer head is too close to the bell it will chatter, if too far off the blow will lose in force and loudness, so either the stop must be adjustable or the hammer stem flexible enough to bend for adjustment.

A bell sounds its best when held loosely. Wood or leather washers of small diameter are the best to hold them, and their screws should not be too tight. Good clock bells are brittle, hard, and easily broken, and the hammer heads should be preferably of brass, giving a better tone than steel, and less liable to injure the bells.

A gong made of wire in the form of a spiral coil gives a pleasing tone. Soft steel wire sounds only moderately well, so the best are of hardened and tempered steel. Gongs made of flat wire instead of round, and of thick and heavy section, give a fuller and richer tone than light ones. The wire should always be brazed into a block of brass, that should in its turn be screwed to the clock plate or to a " standard." A gong standard is a heavy block of iron or steel set upon an upright rod and bolted through the wooden case bottom. This rod vibrates and adds to the tone of the gong, and the wood case bottom acts further as a sounding board.

A gong hammer should have a very elastic " stop " allowing it to bounce a little. The head should be faced with hard leather for softness of tone, and the point at which the gong should be struck is exactly over the centre of its circular coils.

If a jarring noise accompanies the striking of a gong clock, it generally points to something loose in the fixing of the gong standard or the sounding board. In marble clocks with separate sounding boards, if the corner screws are loose, much unpleasant jarring will result. Or in carriage clocks jarring can often be stopped by inserting thin pieces of cork between the edges of the various glass panels if they happen to be easy in their frames, any looseness of the glasses causing trouble.

Straight metal rods are sometimes used as gongs, especially in some German bracket clocks, and do not sound bad, though their tone is uncertain and the sound has no great carrying power. In some of them the " harmonics " are so loud that they actually predominate and it is difficult to say what is really the ground note that the gong is intended to sound. This is especially noticeable when a set of four such rod-gongs is used to chime the quarters. Then the " tune " is sometimes hardly recognizable, some rods sounding their ground note and others their harmonics, which are of course all higher notes.

CHAPTER X.

QUARTER STRIKING AND CHIMING CLOCKS.

Ting-Tang Quarters.—What are known as ting-tang quarters are struck on two bells or gongs, at the first, second, and third quarters, and at the hour, the hour alone is struck upon the lowest. In small clocks the hours and quarters are generally operated from one striking train, and there are two main systems on which it is done, both using the rack form of striking.

In the first system, which may be seen in either English bracket clocks, French carriage clocks, or German clocks, the hours and quarters are all spaced and cut upon one large snail and only one rack is used. This is perhaps the simplest form of all, but has the drawback that it cannot be used in repeating clocks without risk of stoppage or incorrect repeating, or unless a good deal of complication is added. But for clocks not required to repeat, it is the best.

Fig. 87 shows such a snail. It is divided into twelve main sections, one for each hour. Each hour portion is again

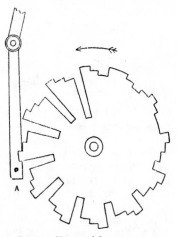

FIG. 87.—Hours and Quarters cut on one Snail.

divided into four. The first small division allows the rack to fall one tooth to strike the first quarter at 15 minutes past the hour. The second small division lets the rack fall two teeth for the half hour. The third lets it fall three teeth for the third quarter. Then comes the fourth section with the hour slot.

There are two hammers, the arbors of which are one above the other, as in Fig. 88, A and B. C is the pin wheel which lifts them both as it turns round. A and B are so spaced that A falls from its lifting pin just before B and so produces the "ting-tang." At the first three quarters both A and B are lifted by the pin wheel. At the first quarter the pin wheel turns one pin space and one "ting-tang" is sounded. At the half-hour the pin wheel turns two pin spaces and "ting-tang, ting-

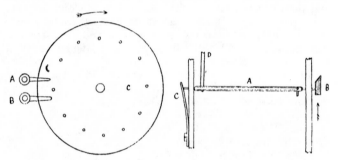

FIG. 88.—Ting-tang Twin Hammer   FIG. 89.—Shifting one Hammer Arbor.
Arbors.

tang" is sounded. At the third quarter the pin wheel moves three pin spaces and "ting-tang, ting-tang, ting-tang" is sounded. At the hour the pin wheel turns as many spaces as the hour requires, and if the hammers remained as before, would sound one "ting-tang" for each stroke of the hour. But there is an arrangement by which at the top hammer A, which hits the smaller gong, slides out of reach of the pin wheel pins and leaves only the hammer B to strike the hour in the ordinary way. Fig. 89 shows the arrangement. There is a cam on the cannon pinion that moves a lever, the end of which, B, is bevelled, and forces the hammer arbor A back about $\frac{1}{16}$ in. so that its lifting arm is out of reach of the pin wheel pins. When the cam on the cannon pinion turns past the hour, the lever B falls down again and allows the spring C, on the back plate, to slide the hammer arbor A back in position again. The arbor B is made with long pivots and has sufficient play between the plates to slide in and out of action. The spring C keeps it in action, forward against the front plate, ready to strike the quarters, and the lever end B presses it back against the back plate out of action at the hour.

No special directions are needed to put such a clock together. As usual, see that the hammer (in this case *both* hammers) are not " on the rise" when the striking train is stopped. In putting on the cannon pinion and cam, see that the lever B comes into action at the right time, and in putting on the snail take great care that the rack tail pin A, Fig. 87, falls freely to the bottom of each hour slot without binding. Clocks made on this plan should not " warn " much before the hour because of the narrowness of the slots in the snail, and are best made with no warning at all, on the plan of Fig. 83.

Another system is that often seen in French carriage clocks and is a modification of the striking work shown in Fig. 83. It has the hour rack, the letting off-lever HJ, the rack detent F, the star wheel, hour snail and jumper exactly as there described. In addition it has a separate quarter snail upon the cannon pinion and a quarter rack which falls upon it. The quarter rack is on the same stud as the hour rack and just behind it. The same gathering pallet gathers both racks together, and the same detent F falls into and holds both racks.

The lever KL (Fig. 83) is in position to keep the hour rack from falling except at the hour, when a pin in the cannon pinion holds it down and lets the hour rack fall.

As in the form first described (Fig. 88), there are two hammer arbors both operated by one pin wheel. Each hammer arbor comes through the front plate and carries a hook or stop A and B (Fig. 90). These stops are held back by a steel arm C, pivoted upon a stud at D. C is normally held down by a spring, but after striking, a pin in the hour rack raises it to the position shown in Fig. 90, holding back both hammers.

As each quarter comes round, the hour rack falls one tooth on to the top of K (Fig. 83), and the quarter rack falls upon the quarter snail. The gathering pallet comes round and gathers the hour rack up one tooth, but both hammers being stopped, a sort of dummy blow is struck.   Then C rises to the position shown in Fig. 91, leaving both hammers free, and the quarters are struck.

At the hour, when the hour rack falls, C falls to the position in Fig. 92, leaving only the lower hammer free to strike, and when the rack is gathered up, C rises again to the position of Fig. 90, the final tooth of the rack only resulting in a dummy blow.

The action is peculiar and a little confusing, but with the help of Figs. 90, 91, and 92, the correct positions can be seen. When taking the clock to pieces, the stops A and B should be

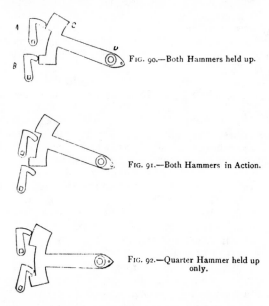

FIG. 90.—Both Hammers held up.

FIG. 91.—Both Hammers in Action.

FIG. 92.—Quarter Hammer held up only.

marked so that they can be put on the right arbors again. Also the arbors A and B should be marked.   Confusing them will cause several mysterious faults.

The " dummy " blow before the quarters and after striking is also confusing, but is necessary and quite correct.

There are other methods of quarter striking, but all depend

upon the suppressing of one hammer when the hour alone is to be struck and restoring it again for the quarters. In all, the two hammer arbors are worked from one pin wheel as here described.

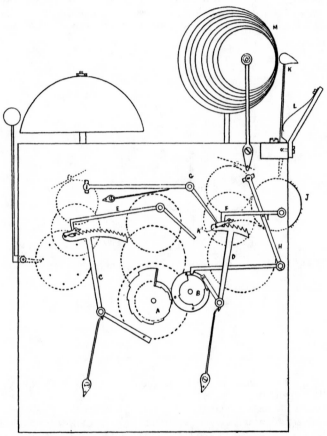

FIG. 93.—English Three-Train Quarter Chiming Mechanism.

Full Quarter Chimes.—Clocks that chime the quarters on four or eight bells or gongs, chiming the four quarters just before striking the hour, require three trains. Generally the centre train is the "going," that on the right is the "chiming," and that on the left the "striking."

Fig. 93 shows the usual arrangement of an English quarter-chiming clock. It will be seen that there are two snails, two racks, two gathering pallets, and two warning levers, etc. The chime bells M are mounted upon a spindle supported by two

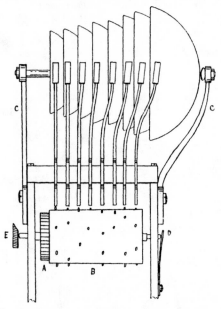

Fig. 94.—Arrangement of Pin-barrel, Hammers, and Bells.

uprights screwed to the back and front plates respectively. The row of hammers, K, are jointed into a brass block screwed across the top right-hand corner of the frame and are supplied with light springs, L, to cause them to hit the bells when drawn back and released. The hammer tails are drawn back by pins in a chime drum or pin barrel, musical-box fashion, marked J. The pin barrel, hammers and bells are shown in Fig. 94. In this figure, B is the pin barrel, A the wheel mounted upon it, by means of which it is driven from the second wheel of the chiming train, CC are the bell standards or supports screwed to the frame plates.

The circumference of the pin barrel is divided into five portions, in each is pinned a "chime." The barrel revolves twice per hour and for each tooth of the $\frac{1}{4}$ rack gathered up, that is, for each revolution of the pallet wheel, one portion or section of the pin barrel passes under the hammer tails and one "chime" is struck. Thus, at the first quarter, the rack falls upon the first step of the $\frac{1}{4}$ snail B (Fig. 93), one tooth is gathered up, and one chime struck. At the half-hour, the rack falls two teeth and two chimes are struck. At three-quarter hour, the rack falls three teeth and three chimes are struck. At the hour it falls four teeth and four chimes are struck. The sum of these, 1, 2, 3 and 4 = 10. As there are five chimes pinned on the barrel, that equals two revolutions. No. 1 chime is given at the first quarter, Nos. 2 and 3 at the half-hour, Nos. 4, 5, and 1 at three-quarter hour, and Nos. 2, 3, 4, and 5 at the hour.

The chime bells, generally eight in number, are all screwed upon one long steel spindle, separated by small wooden washers. As a rule these clocks are arranged to chime on either eight or four bells, at pleasure. "Westminster" chimes are on four bells and are taken on the 3rd, 4th, 5th, and 8th bells, No. 1 being the smallest and No. 8 the largest. The order of the chimes is as follows :—

```
Chime No. 1— 3, 4, 5, 8
    „       2—5, 3, 4, 8
    „       3—5, 4, 3, 5
    „       4—3, 5, 4, 8
    „       5—8, 4, 3, 5
```

These are the chimes now almost universally used when four bells are available, both in house and turret clocks.

When eight or ten bells are used, there is no fixed order of chiming, but two rules are nearly always observed, first, the No. 1 chime is a simple "run," thus :—

$$1 = 1, 2, 3, 4, 5, 6, 7, 8, 9, 10$$

Second.—Each chime ends on the lowest bell of the peal. Beyond these two rules, the fancy of the maker may have full play, and among the many used, the following may be given as favourites for eight-bell chimes :—

$$2 = 4, 6, 5, 3, 7, 2, 1, 8$$
$$3 = 1, 3, 5, 7, 2, 4, 6, 8$$
$$4 = 1, 4, 3, 6, 5, 7, 2, 8$$
$$5 = 4, 3, 2, 1, 5, 6, 7, 8$$

The "change of chimes" is effected by sliding the pin barrel about $\frac{1}{16}$ in. A flat spring D (Fig. 94) screwed to the back plate presses upon the elongated pivot of the barrel axis and keeps it forward. In this position the circles of pins used for the eight-bell chimes come under the hammer tails. A lever with a bevelled edge E (Fig. 94) is used to force the pin barrel back so that the circles of "Westminster" pins come under the hammer tails. The end of the lever E may be arranged to be operated from the dial edge or from a small circle in the "arch" of the dial, as fancy dictates.

Returning to Fig. 93, the hour snail A is mounted upon a star wheel moved by a pin in the cannon pinion, as described and shown in Fig. 83 or upon the hour wheel. The hour rack C is of the usual pattern and has its spring and gathering pallet. The hour rack hook E has a long tail. The warning lever G also has a long tail, and is kept up by a spring. When up it stops the warning pin. The quarter snail B is on the minute wheel, and has at its back four lifting pins which raise the quarter warning lever H. A pin, I, halfway up H raises the quarter rack hook F, and lets the quarter rack D fall.

The mechanism acts thus: As the first quarter is approached, a pin in the minute wheel lifts the lower arm of H until the pin I lifts the rack hook F and the rack D falls until its tail rests on the first step of the quarter snail. The upper arm of H arrests the warning pin as in an ordinary striking clock. The pin in the minute wheel, continuing to advance, lets H fall just at the quarter, liberating the warning pin and the rack hook, and the chiming train runs one revolution of the pallet wheel, while the rack is gathered up. This is sufficient for one chime to be struck.

At the half-hour and at three-quarter hour, the same thing occurs again, the rack falling two and three teeth respectively. At the hour, the minute wheel pin comes round as before and lifts H, the quarter rack falls four teeth and its left-hand end strikes the tail of the hour rack hook E. This liberates the hour rack C, which falls, and the striking and chiming trains both run until their warning wheels are stopped by the warning levers H and G. G is automatically raised by its spring as soon as the quarter rack falls each quarter, but at the first three quarters the striking train is locked by the wing of its gathering pallet upon the rack pin. But at the hour, the rack C having fallen, G catches the warning pin. Just at 60 minutes, H falls and the four chimes are struck. As the last tooth of the rack D is gathered up, the pin K pulls down the warning lever G and liberates the striking train. The hour is then struck.

To thoroughly clean and repair one of these clocks entails a great deal of work. The process is the same as that used in dealing with an ordinary grandfather clock or eight-day dial, but the workmanship being as a rule better, greater care may be taken in regard to polishing. It is best when taking the clock to pieces to keep the striking and chiming trains separate, as the pallet wheels, warning wheels and flys are so much alike.

When putting together, the same points must be observed as in ordinary striking clocks. The hour hammer must not be " on the rise," and the warning pin must have half a turn of " run." In the chiming train the pin barrel J must stop between two chimes exactly, with no chime hammers half lifted. As a rule J is made to take out separately and can be easily placed correctly, especially if the gathering pallet be placed on the proper square. But the chiming train warning wheel will require adjusting like that in the striking train. These adjustments all have to be made by raising the plate and liberating one pivot of the wheel to be moved, turning the wheel round, and replacing. It is an awkward method and takes time, but must not be left incorrect or trouble will surely result.

In a quarter-chiming bracket clock with fusee and chains, some little difficulty may be experienced in putting on the going fusee chain, and it should therefore be put on first, before the other two make it more awkward still. The plain fact is there is room for improvement in the arrangement of these clocks; as at present constructed there are too many wheels between one pair of plates.

All lifting pins should be smooth and oiled and the

I

friction and resistance of letting off the quarters lessened as much as possible, or it may seriously diminish the power of the clock. The cannon pinion should be firmly gripped by its spring so that it may turn stiffly enough to carry under all circumstances. If at all easy to turn, the resistance at the quarters will cause it to lag behind.

It should always be remembered that the more work a clock is required to perform, the less free it is to keep even time. Letting off quarter-chiming work, moving date work and moon discs all take from the power available for driving the train and pendulum. It must happen occasionally that *all* these movements take place at the same time, and this necessitates a driving power (if the clock is not to stop) far in excess of that required for merely going and showing the time. From this it follows that these clocks must generally have more power than they require to provide against the occasional moments when it is reduced by one-half, and the result is excessive wear and tear and undue pressure on pivots, etc.

Locking-plate Chiming Mechanism.—A quarter-chiming clock could be equally well made upon the locking-plate principle, having one locking plate for the hour striking train and one for the quarters. Or a two train "ting-tang" clock might be made with all the hours and quarters set out round one large locking plate. But they would both be liable to the fault, viz. if the striking failed at any time, the entire series would be upset. The same thing would occur if the clock were allowed to run down, whereas with rack striking and chiming, if one hour or quarter failed, the next would strike correctly provided nothing jambed in the train or snails, and the best rack clocks are provided with spring tails to the rack so that if failure occurs, the tail will rise and pass over the snail without injury.

In turret clocks, which are attended to more regularly than house clocks, and generally by an expert clockmaker, locking-plate chiming work is the rule and racks the exception. The work being larger and heavier is also much less liable to failure and no trouble arises.

Tube Chime Clocks.—Quarter-chiming clocks, chiming on long tubular gongs, differ in many respects from bell clocks.

The tubes being large and heavy require heavy hammers to bring out their tone properly, and as a matter of fact, no house clocks have yet been made that have hammers sufficiently heavy to get the best out of the tubes. To raise heavy hammers requires a great deal more driving weight than ordinary clocks are provided with, and also stronger wheelwork. Whereas an eight-bell chiming grandfather clock will go with a weight of 20 lbs. upon the chiming side, a tube clock with the smallest tubes and lightest hammers possible requires 36 lbs. and could do with 50 lbs. easily. When wheels, pinions, and barrels of ordinary size and thickness, as used in bell clocks, are used with a driving weight of this amount, the wear and tear is far too great and the friction harsh. The surfaces in contact are not large enough to stand the pressure. There is not room upon the barrel for a line strong enough to carry the weight properly.

Such clocks always ought to be made with extra long and large barrels, as long as can possibly be got between the plates, say another ¼ in. long and as large in diameter as the height of the case will allow for the fall of a weight upon a double line. For increasing the fall is equivalent to increasing the weight and adds no extra strain upon the line. So tube clocks are, as a rule, made higher than usual, with deep frames, an extra large chime barrel and a chime mainwheel (where the most strain comes) of large diameter and thickness. The chime weight is suspended upon a wire line to lessen the chance of breakage.

Still, with all these concessions, the arrangement is far from ideal. The weight is still too light to raise hammers of the proper weight to bring out the full tone of the gongs, and nothing short of a separate movement for the chiming mechanism will do it, but this presents difficulties, and the result is a compromise as we see it to-day.

Another weak point in these clocks is the change in direction of motion necessary to operate a pin barrel that extends across the back of the movement. This is done sometimes by bevel wheels and sometimes by a contrate wheel or crown wheel. Fig. 95 gives an idea of the manner in which the pin barrel and hammers are arranged for the chimes, and also the hour hammer. The pin barrel A is supported by brackets so as to revolve at least one inch clear of the back plate, and generally more. This is to allow room for the pendulum and the large " regulating " flys. In Fig. 95, the pin barrel is driven by the crown wheel D mounted on the axis of the second wheel of the chiming train. D drives a wheel E mounted upon the circumference of the pin barrel. But in perhaps the majority of clocks bevel wheels are used at the end of the barrel. The crown wheel method gives less shake between the teeth and wastes less power. In clockwork it is awkward to mount bevel wheels and ensure good depths, hence the waste of power.

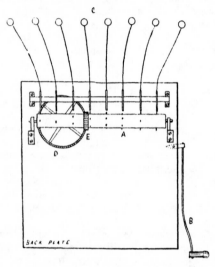

FIG. 95.—Pin-barrel and Hammers of Tube Chime Clock (Back View).

Theoretically, of course, properly cut bevel wheels are far the best, but practical difficulties nullify their advantages, and clocks driven as Fig. 95 will go with a less weight, which is proof of greater efficiency.

The hammers are long, heavy, and springy, and are supplied with light springs. They are spread and bent backward so as to reach the upper ends of the eight tubes. Fig. 96 shows the hammer and tubes in position, and Fig. 97 the manner of suspension of the tubes. G is a hard wood rail going across the top of the case back. From it the tubes H are suspended by cords passing through holes in their sides. The hammers, their faces softened by a covering of kid or thin rubber cloth, strike the tubes from ½ in. to 1 in. from their top ends. The hour tube is hung from a separate bracket either just above or just below the seat board, and the hour hammer, B, hangs downwards. F are the hammer springs.

The same directions apply to cleaning and repairing these clocks as to bell clocks, the only special points relate to the

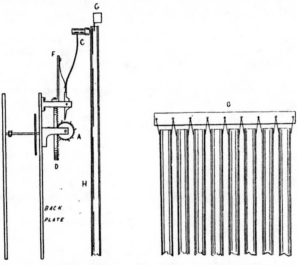

FIG. 96.—Pin-barrel and Hammers of Tube Chime Clock (Side View).

FIG. 97.—Suspension of 8 Quartet-tubes.

hammers and tubes. If the hammer coverings have worn thin or hard, the tone will suffer and they should be renewed. Attention should also be directed to silencing the hammer stems. If they are arrested or checked metal to metal, a rattling is heard during chiming, and it is best to insert leather or thin rubber pads for the hammers to fall upon. These also have a tendency to get hard and frequent renewal is advisable. Equality of tone during chiming can only be obtained by careful adjustment of the hammer stems to strike as hard as possible without chattering, and by giving the springs equal strength. If the springs are set on too hard, the clock will fail to lift the hammers, so care has to be taken not to give the clock too much to do, but at the same time to get as much work as possible out of it.

The tubes being generally nickel plated, should be rubbed down with a leather, or if tarnished, polished with "Globe" polish and a rag.

On account of the heavy pressure, the chiming train of a tube clock will want oiling more frequently than an ordinary clock, or the pivots, etc., will suffer.

Musical Clocks.—Clocks that play tunes periodically on bells or tubes have a large pin barrel that makes one revolution for each tune. As a rule the locking plate system is used, the

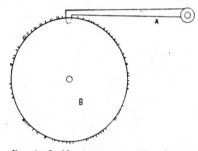

barrel being its own locking plate, as in Fig. 98. The locking arm A enters a notch cut in the end of the pin barrel B, and falling, stops the train running. When raised, the train starts and runs while A rests on the circumference of the barrel until it again drops into its

FIG. 98.—Locking Arrangement of Tune Clock.

notch. The whole arrangement is similar to an English locking-plate clock as in Fig. 77, with lifting arm, warning lever, etc., complete.

These tune-playing attachments are sometimes seen in grandfather clocks, playing the tunes on bells, sometimes in German and Swiss clocks playing on a kind of musical-box comb, and occasionally in an old German clock, playing on a set of organ pipes.

When the latter are used, the train has also to work the bellows, and the clock has a windchest, a set of valves opened by levers actuated by the pins in the pin barrel, and a "soundboard" upon which the pipes are mounted or have their wind conducted to them by tubes. The whole arrangement constitutes a miniature pipe organ and is subject to all the little faults and troubles of organs. As nearly all of them are old, putting one in thorough order will generally prove a work of time and patience.

Changes of tune are sometimes automatic and sometimes effected by a hand lever, and any number of tunes from one to ten may be pinned upon the barrel.

Among the possible faults may be mentioned rattling caused by worn pivots and wide pivot holes; bent pins in the pin-barrel; leaky bellows; valves not closing air-tight and admitting air to pipes when not wanted; leakage of air from one division to another in the soundboard caused by warping of the wood, or perished leather; glue joints that have started; and pipes that have parted at the mitred corners, for in a small compass it is necessary to " mitre " the longest pipes several times and practically wind them all round the clock. The ravages of " worms " have not been mentioned, but are by no means negligible.

Cuckoo Clocks.—These are of the wood-frame variety and simply locking-plate striking clocks with two hammer arbors like a " ting-tang " quarter clock. As the hour strikes, the principal hammer hits the wire gong and the twin hammer arbors raise two wires which lift the little bellows at the top of the clock and let them fall again one after the other. Each bellows gives a puff of air to an organ pipe let into the case side, producing the " cuckoo." So each stroke of the hour consists of a blow on the gong and a " cuckoo." The bird is a little mechanical toy that when its tail is raised by a wire on the top of one bellows, bends forward, opens its beak and wings, and apparently gives a " cuckoo."

These and all " Dutch " clocks are simplified by the fact that each train is held by a separate lath of wood. Withdrawing the pins at the top enables either lath to be taken out and the train removed. Great care must be taken to see tha no lifting arbors are " on the rise " when striking ceases. Also that the cam wheel coincides with the locking plate, or the striking will not act properly. The warning also must be attended to—in fact, all these points are as essential to the proper going of a cuckoo clock as to an English bracket clock, and if not made perfect, failure will result.

The striking train really contains no new feature, and if the chapter on striking clocks is carefully read, cuckoo clocks will only need patience to do correctly.

The novel features in these clocks are merely in points of construction. Chains take the place of lines; wood serves instead of brass for frame plates and sometimes for the mountings of wheels and pinions; the pivot holes are brass bushes let into the wood, and the pendulum is generally hung on a wire loop instead of a spring, which wire loop must be well oiled or it will continually squeak as the clock goes.

Ship Clocks.—One other kind of striking clock deserves notice, and that is the ship clock that strikes the " bells " used to denote the divisions of time on board ship. These strike every half-hour from one to eight " bells." One bell is a single stroke ; two bells a pair of strokes in quick succession ; three bells, one pair in quick succession and an odd one, thus : "ting-ting, ting," and so on. No special directions will be required for dealing with them, it being quite simple. These clocks are usually of the American lever drum pattern, about 6 or 7 in. diameter, or else like a small eight-day English dial with a large carriage clock escapement. Pendulums of course cannot be used. Also exact timekeeping is not necessary as they are continually being altered as the ship's position varies and the time of local noon changes.

CHAPTER XI.

TURRET CLOCKS.

UNDER this heading come all clocks showing the time on out-side dials or striking or chiming on bells mounted in belfries so as to be heard outside the building in which they are placed. They range in size from that used outside a watch-maker's shop and the clock mounted in a small turret over a stable in a gentleman's house, to those built in large towers like that at Westminster, having four dials and chiming on a heavy peal of bells.

The size and weight of such clocks and the conditions under which they work render a different system of construction necessary. They may be roughly classified in two groups— ancient and modern. The dividing line is a sharp one and dates from the construction of the Westminster clock to the design of the late Lord Grimthorpe, who though a lawyer, was the most advanced clock designer of his day. This clock was a radical departure from the previous practice in almost every detail of its construction, and has scarcely had an improvement made upon it to this day.

The "ancient" turret clock had a frame of the "birdcage" pattern, generally of wrought iron, containing the wheels, etc. Bars were bolted across this frame to hold the separate trains, and the whole clock generally rested upon a timber support, or a couple of small beams let into the walls. The escapement was generally a "dead-beat" with steel pallets, like Fig. 63, or occasionally a recoil like Fig. 23. In the largest, a pin wheel escapement, like Fig. 99, is sometimes seen, and occasionally "freak" escapements were met with. In Fig. 99, B is the escape wheel, having round its rim a series of hard steel pins which fall upon the dead faces of the pallets A. This is a

good escapement of its kind and especially calculated to with-
stand wear and heavy pressure.

In these old clocks, the pressure upon the escapement was
very severe, as a heavy driving weight was necessary to drive
large outside hands. Strong winds and snow have a very
powerful dragging effect and will some-
times stop the hands of the most
powerful clocks. To overcome this
resistance a heavy weight is absolutely
necessary and the whole of the
pressure comes upon the escape wheel,
teeth and pallets.

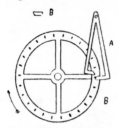

FIG. 99.—Turret Clock Pin-
wheel Escapement.

In some old church clocks with
gun-metal escape wheels with straight-
pointed teeth like Fig. 63, the pressure and hammer-like blows of the
"drop" on the pallets have gradually
bent all the teeth in a curve.

The striking of these clocks is sometimes by the "rack"
system, but more often by "locking plate." Many of the older
ones have wooden barrels with steel spindles, like mangle
rollers, and are provided with hemp ropes for the weights.
Hardly any two clocks are found exactly alike or even on
quite the same plan, and seeing that hundreds of them were
the work of local blacksmiths, they are marvels of good honest
workmanship.

The pendulums varied from 14 ft. (beating two seconds) to
39 in. and sometimes weighed several hundred-weights. Wood
rods were the rule, though many iron ones are seen.

Modern turret clocks have frames of "bedstead" or
horizontal pattern, to which each wheel is held by separate
gun-metal bushes bolted on, each being separately removable.
The horizontal bed frame is a single iron casting and generally
is supported upon a pair of cast-iron brackets bolted to the
stone walls 'of the clock room. The mainwheels are of cast
iron and the rest of gun-metal. The lines are flexible woven
steel wire. The pendulum is a "compensation" of the zinc and
steel variety, or in the latest, of "invar." But the greatest of
all improvements, the escapement, is a "gravity," by which the
great power of the driving weight is absorbed before it reaches
the pendulum, the latter being driven by a pair of gravity arms

raised by the train at each beat, but falling on the pendulum rod by their own weight only. Obviously it matters not how much power is expended upon raising them, whether excessive or only just sufficient, so long as they are raised. The weight in falling is constant, and thus the pendulum swings through a constant arc, no matter how much the power varies according to the weather, wind, snow, etc.

Fig. 100 shows the " double three-legged" form of gravity escapement generally used in large clocks. There are two escape wheels mounted on one axis about $\frac{1}{2}$ in. to 1 in. apart. Between them are three small lifting pins near the centre, forming a kind of " lantern pinion." The duty of these pins is to lift the gravity arms free from the pendulum, the arms of the escape wheel being locked by one of the locking blocks S.S'. The escapement acts thus : the pendulum rod travelling to the right moves the right gravity arm until the escape wheel is liberated from the stop S. The lower centre pin then lifts the left-hand gravity arm clear of the pendulum rod and the wheel is locked again by the arm (small) C, resting on the stop S'. The effect of this is that the weight of the right-hand gravity arm rests against the pendulum rod and drives it onward. On the return swing, the right-hand arm is lifted by the escape wheel, and the left one does the driving, and so on.

Fig. 100 is reproduced from " Clocks, Watches, and Bells," by Lord Grimthorpe, the inventor of the escapement. Another form of the same escapement, used more in smaller clocks, is the " single four-armed gravity," also invented by him. This is on exactly the same principle, but has only one escape wheel, provided with a double set of four lifting pins.

The effect of the adoption of these escapements and compensation pendulums in turret clocks has been to reduce their errors from minutes to seconds per week, and cannot be over-estimated.

A few large turret clocks have no hands or dials at all except a small " indicator" dial inside the clock room, like that at Louth in Lincolnshire, which chimes Westminster chimes and strikes on a heavy peal of eight bells in the church tower, but gives no other outside indication of the time. The architectural design of some towers precludes the use of outside dials without disfigurement.

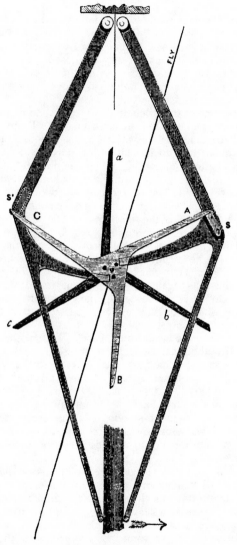

FIG. 100.--Lord Grimthorpe's Double Three-legged Gravity Escapement.

Turret clocks, especially modern ones, when regularly wound by a clockmaker, and given just a little constant attention, need never be taken to pieces except for repair. They can be kept clean like other running machinery by oiling freely from time to time and wiping off all surplus, brushing through the teeth and pinion leaves with petrol and wiping dry, etc., as they lie. The motion is slow enough to allow of this, and in the case of chiming and striking trains, they stand still for periods of at least ten minutes together and give excellent opportunites for thorough cleaning.

The escapements and lighter wheels should be oiled with light machine oil, the mainwheels and striking trains with a heavier oil, and the hammer-lifting cams on the main wheels should be greased with lard. The beat pins at the ends of the gravity arms, that rest on the pendulum rod, should be clean and dry. Stop blocks require just a trace of oil, and of course all the escapement pivots and the centre lifting pins.

Steel lines should be thickly greased or vaselined to prevent rusting. This also avoids friction between the coils as they are wound on the barrels. All pulleys and cranks of the bell-hammer work and driving lines need constant attention, in the way of oiling and cleaning.

When a turret clock has been neglected for a long time and only been oiled a little from time to time, as when in the care of a gardener, verger, or caretaker, as many small clocks are, the time arrives when it must be taken apart, cleaned and overhauled.

The first thing to do is to let down the weights and relieve the clock of all power, or a serious accident may result. It generally requires two to do this, one to hold up the winding clicks and another at the winding handle. Even then, with a heavy weight there is some danger of the handle flying round and doing damage, so it is best when possible to let the clock run down first. Having let down the weights, the trains may be taken out and the bed plate or frame cleaned thoroughly with petrol and scraped. The parts so treated should be taken out into the light and air, and on no account must petrol be freely used in a small dark clock room in a tower, working by the light of a candle or lamp, or an explosion may easily occur. Parts like the bed frame that cannot be removed, when in a dark place, are best cleaned with paraffin first, then scraped, and washed with hot soda and water.

When the main wheels are of cast iron, the teeth may be greased with lard, but from time to time it must be wiped off or a dirty mess results. Gun-metal wheels should be left clean and dry, and also the pinions they drive. Steel arbors and levers may be wiped finally with a paraffin rag to make them rust proof. The motion work that leads to the hands should always be freely oiled, as damp gets in here very badly from the outside of the dial.

If it is necessary to shift the fly or warning wheel when power is on the clock, first securely tie up the next wheel with rope to prevent it running, and do not trust to inserting a screwdriver between the arms, as such a proceeding invites accident.

The old rule that the hour hammer must not be " on the rise " does not apply to turret clocks, as there is ample power, and indeed in good clocks it is the rule to make the striking train stop with the hour hammer lifted up almost ready to fall. This is to ensure promptness in striking, so that the first blow may be relied upon to fall almost exactly to time, otherwise the train starting slowly might take a variable time to raise the hammer for the first blow, depending on the temperature and the thickness of the oil, etc., and render the time of striking uncertain to a few seconds. But, as in house clocks, all warnings, etc., must have a clear run.

Hemp ropes in old clocks are generally nailed to the wooden barrels, and frequently require renewing. When in conjunction with iron barrels it is much the best plan to replace them with flexible steel lines. Steel lines if put on wooden barrels cut the wood to pieces, but if the wood is protected by a sheath of thin sheet iron nailed around it, steel lines may be put on these also with advantage.

The breakage of ropes is one of the chief troubles in connection with small turret clocks of the older kind as seen in village churches, etc. They and their pulleys are often in the most inaccessible places. Sometimes the lines are led from the barrels upwards to a top pulley in a corner near the leads of the roof to give the weight a straight fall to the floor, at other times, the line is led to a corner from beneath the clock and the weight falls down a shoot through the church and continues down a " pit " for another ten or fifteen feet into the earth.   When the line leads to the roof and the weight falls to the floor of the clock room over the church entrance, care must be taken to have a bin of sawdust or gravel under it to break the fall when a line gives way, or it may go through the floor.

When a line breaks and lets the weights fall into its " pit," it is often troublesome to fish the weight up again, the pit being generally far too narrow for any one to go down.   It can generally be managed by lowering a candle or lamp and fishing with an iron hook and the remains of the broken line.

As far as possible turret clock movements should be boxed in free from damp (which often drips from the ceiling of clock rooms and runs down the lines) and from dust and birds. When lines pass upwards from the clock, they should go through a hole with a cloth washer, in a lath, designed to slide along over a slit in the case top, and so let no wet in.

CHAPTER XII.

MAKING CLOCKS.

THE hand-made clock industry of England is still a large one, and consists mainly of eight-day shop dials, bracket, striking, and chiming clocks, grandfather clocks, long-case quarter-chiming and tube clocks, skeletons, a few carriage clocks, ship clocks, regulators for workshop or observatory, and turret clocks.

For these and all similar clocks, parts in nearly every stage of finish can be purchased in the clockmaking centres such as Clerkenwell or Birmingham. Brass castings for skeleton plates, cocks of all sorts, for minute wheels, flys, pendulums, etc., bridges, and clicks; steel forgings for pallets, striking racks, chime hammers, and bell standards; stamped brass wheel blanks; rough wheels with the teeth cut, in sets for the various clocks; German cut pinions; pinion wire; rough barrels and fusees with the spiral grooves turned; sets of chime bells, gongs and tubes, and almost every detail wanted by the amateur or professional maker.

To give an idea of the condition of the various parts, and of the amount of work necessary to convert them into a going clock, a few photographs are given. The parts or "rough materials" shown were photographed just as they came from their various makers, and are exactly as a Clerkenwell clock-maker would purchase them if he were given an order for a clock of that description.

Fig. 101 shows various rough wheels. A is a blank hour wheel, cut from sheet metal, rough filed on both sides, and with the teeth accurately and smoothly cut by a wheel-cutting engine. B and C are escape wheels, stamped from sheet metal, with the teeth engine-cut. D is a main wheel for a grandfather clock and is cast, the flat sides only being rough-filed and the teeth engine-cut.

It may be here stated that all clock wheels, whether made from sheet metal by turning, stamped out by dies, or cast, have the teeth cut by a wheel-cutting engine, and in a rough wheel as shown here the teeth are the only parts "finished."

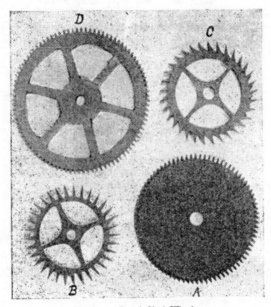

FIG. 101.—Rough Clock Wheels.

It is an error to suppose that even in the most common clock the teeth are stamped out.

Fig. 102 shows various rough pinions. A is a length of pinion-wire. This soft steel wire drawn through a special draw plate to the section of a pinion with six, seven, or eight leaves as the case may be. This wire can be obtained in any size and in lengths of one foot. To make a pinion from it, a short length like that illustrated is cut off, the leaves are filed off where not wanted, the wire is centred in a lathe and rough-turned true, then hardened and tempered, the leaves smoothed

K

and polished and the arbor and pivots turned. B is a German pinion as imported for the use of clockmakers. Its leaves are machine-cut, it is ready hardened and tempered, is centred and rough-turned true, and the leaves are polished. B has twelve leaves and is rough "regulator pinion." C is a similar German pinion of seven leaves as used in an eight-day dial. D and E are "finished" German pinions ready to have the wheels mounted upon them and run in the clock frame. D is an escape pinion, and E a centre pinion for an eight-day dial. F and G are "lantern" pinions, not purchasable, but made by

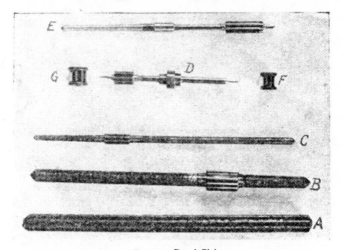

FIG. 102.—Rough Pinions.

hand by turning a brass "bobbin," drilling the holes, and inserting steel "leaves." These are more often made by amateurs than by clockmakers. They are used by driving them on to slightly tapered arbors and are included in this photograph merely to show the various kinds of pinions in use.

Fig. 103 shows several rough forgings as bought from the material shop. A is a rack-hook, B is a rack, C is a pair of "recoil" pallets, D is a pair of "dead-beat" regulator pallets, and E is a gathering pallet. Forgings for many other parts are obtainable, these being only shown as samples.

Fig. 104 is a photograph of the little heap of rough materials required to make the skeleton clock shown in Fig. 1 of this book. In it may be seen the castings for the pair of plates, pendulum cock, minute cock, barrel click, and four feet for the frame. The three finished pinions, the chain, the mainspring in its wire, the rough-turned barrel and fusee, and the cut blanks for main wheel, centre, third, escape wheel hour wheel, minute wheels and ratchets.

Fig. 105 shows a set of wheels and pinions for a watch-

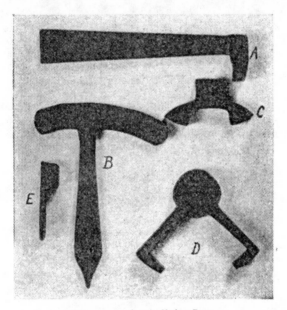

FIG. 103.—Forgings for Various Parts.

maker's regulator clock. The going barrel will be seen with its spiral groove turned and the ratchet cast on and cut, all in one piece. The arbor is simply rough steel driven through, centred in the lathe. The three rough pinions are German, of twelve leaves. The main wheel is cast, and the others are cut from sheet brass.

A few directions for the practical working up of such materials will now be given.

Suppose it is desired to make a clock on the lines of an eight-day dial like that shown in Fig. 53. Two pieces of

FIG. 104.—Rough Materials for making a Skeleton Clock.

sheet brass will be required for the plates about $\frac{1}{8}$ in. thick. At the metal merchants' these will be sawn out to the required dimensions and weighed up at so much per lb. It will be advisable to let them "planish" them. As sawn out from

FIG. 105.—Rough Materials for making a Regulator.

rolled sheet, they will be far from flat and by no means smooth. Planishing is a hammering process by means of which they are flattened and smoothed almost like glass. At the same time they are slightly hardened. It costs little and saves much labour.

Making a Frame.—The plates being sawn out and planished, a centre line is drawn down each by a fine steel point, guided by a steel "straight-edge" as a ruler. On this centre line, two holes are drilled about $\frac{1}{16}$ in. diameter, near the top and bottom of each plate, and the plates are then riveted together with two brass rivets through these holes. Held thus they are filed both together square and true on the edges. The pillar holes at the corners, or wherever they are required, are drilled and opened out to size, about $\frac{1}{4}$ in. diameter. Then the plates may be separated and the pillars may be made.

Castings for these may be used if desired, but as each casting requires centring separately in the lathe and the rough turning off, it saves labour and produces a harder and sounder pillar, to turn them all from a foot length of $\frac{3}{8}$ in. or $\frac{1}{2}$ in. round brass rod. This brass rod is drawn smooth and true, and is much harder and generally of better quality brass than small pillar castings. Once centred in the lathe, pillar after pillar can be turned from its end, and the cutting-off process makes it easy to drill up a fresh back centre to turn the next.

The pillars are riveted into the back plate very firmly, and come through the front plate about $\frac{3}{16}$ in. to allow of being drilled for the pins.

The barrel and fusee will be purchased in the condition of those shown in Fig. 104. The wheels will be like those in Fig. 101. First take the barrel in hand and file out the hollow in the cover for taking-off purposes. Drill and cut the chain hook hole, or the three holes for the line. The mainspring hook will have to be riveted in. The arbor must have its hook filed up clean and well undercut to hold the eye of the spring. The barrel itself must be smoothed finally in the lathe and polished with pumice powder and water on a rag, or with "Globe" polish. The arbor must have its pivots turned and polished. The frame being made, the correct endshake of the barrel arbor and all the wheels may be gauged by laying the arbors across the edges of the plates.

Trueing Wheels.—The wheels must next be taken in hand and their centre holes turned true. As received from the wheel cutter's there is no guarantee that the centre holes are exactly in the centre. So a piece of smooth-grained wood is bolted or screwed to the lathe face plate and a shallow sink is turned in it, into which the smallest wheel (a minute wheel) can be just pushed tight down to the bottom, held by the points of the teeth. Thus held true, its centre hole can be turned out by a sharp hand or slide-rest cutter, and this ensures the wheel running quite true when mounted upon its pinion.

The smallest wheel done, take the next and enlarge the sink in the wood chuck until it will just push in tight, and so on until all are done, the largest being treated last.

Once turned true, they may have the centre holes enlarged by broaching if required, without affecting their truth, though it is advisable not to leave too much for broaching, but to turn them approximately to the sizes required.

The main wheel must be opened out to go on the fusee arbor, which is turned parallel and polished smooth to receive it. The groove for the "key" is turned, the key made and fitted, and the fusee arbor pivoted to the required endshake.

The pinions are turned, polished, and pivoted, and the wheels smoothed and mounted

FIG. 106.—Riveting a Wheel upon a Pinion.

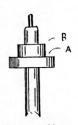

FIG. 107.—Mounting a Wheel upon a Brass Collet.

upon them, laying them across the plate edge to see that they work together and have proper freedom. In clocks, some wheels are riveted upon the leaves of the pinions and some are mounted upon brass collets. Fig. 106 shows how a pinion is turned to receive a wheel. A square shoulder A is turned upon which the wheel can be forced tight. Then the points B are riveted over upon the centre of the wheel to hold it tight and prevent it turning. The face of B is turned hollow or undercut, leaving the points of the leaves standing up a little above the surface of the wheel to facilitate riveting.

Fig. 107 shows how a wheel is riveted upon a collet. The collet is turned from brass and may be driven tight on to a slightly tapered arbor, or bedded with soft solder. The wheel is seated upon the square shoulder A and the edge B, undercut as usual is riveted over its centre as before. To prevent turning round, the inside edge of the centre hole of the wheel is "starred" by cutting a few little notches in it for the brass edge B to be riveted into.

Pitching the Train.—The train must next be " run in " the plates. The centre pinion hole is first marked with a fine centre punch, then the fusee hole below it on the centre line, or on one side of it, according to the design of the movement. To mark the exact position of the fusee hole, the depth must be measured. With a pair of fine dividers from a set of drawing instruments, measure the pitch diameter of the main wheel (see Fig. 20). Rule a fine line on a piece of paper and prick off this distance with the divider points. Then measure the pitch diameter of the centre pinion in the same way and add it to the other distance on the line upon the paper. This gives the sum of the full pitch diameters of the wheel and pinion. Exactly half this distance is the " depth." So with the dividers and the paper, the half is found by trial, stepping it with the divider points. When found quite correctly, transfer it to the plate and mark the position of the fusee hole.

Drill no pivot holes until all the depths are marked with the centre punch. Next mark the position of the escape holes upon the centre line. The position of the third wheel hole must be scored off with compasses to suit the two depths so that its pinion may run with the centre wheel and its wheel may drive the escape pinion. With dividers as before, measure the pitch circle of the centre wheel, mark it on paper, and add the diameter of the third pinion to it. Halve this as before, and using the dividers as compasses, strike an arc of a circle from the centre wheel hole as a centre. Measure the third wheel and escape pinion in the same way, add them together, halve them, and score off another arc of a circle cutting the first one, and using the escape hole as a centre. Where the two arcs intersect is the position for the third pivot hole. Mark it with a centre punch as before.

The escapement is best drawn out very carefully on paper or zinc, like Fig. 23 (see " making pallets "). The distance between escape pivot and pallet pivot may be then transferred to the plate and marked.

The barrel must be placed so that it just clears the main wheel teeth, and comes in a convenient place in the frame, being clear of the pillars. Measure its full diameter and add to it the outside diameter of the main wheel. Halve this and mark the barrel holes a little further off than the mark, say $\frac{1}{8}$ inch for freedom.

Having all the pivot holes marked, take a fine drill much smaller than the pivot holes and drill the plate half through. This is to ensure the full-size drill starting true. A drill of full size might wander from a centre punch mark and not drill the pivot hole just where required, but it will follow a small leading hole quite accurately.

Next pin the plates together again by their top and bottom holes, as when squaring the edges. They need not be actually riveted, but two brass pins driven through tight will answer just as well and can be easily knocked out again. While pinned together, drill the pivot holes straight through both, or through the first and partly into the second, so that they may be completed afterwards, as some drills begin to run hard or choke when in a certain depth, and the two plates together require a rather deep hole for a small drill. The fusee and barrel holes may be drilled with a larger drill and easily put right through.

Separate the plates and open out the holes to fit the pivots accurately, not too large, but just so that the wheels spin freely. The great point to observe in opening out a pivot hole is to keep it upright.

When all are drilled and opened out, run in the wheels, at first each one separately to see that each is free and has proper endshake. Oil the pivots to prevent scratching the polish. Then run them in pairs to test each depth by itself. Finally put in the whole train and test for freedom.

Generally if the dividers have been applied carefully and sharp drills have been used for the pivot holes, all the depths will be perfect. If not, the erring pivot hole must be " drawn " by opening with a broach and filing to one side a little, then bushed.

The back cock or pendulum cock may be screwed on. Then a broach put through the back pivot hole in the plate will mark the pivot hole in the cock, which may be drilled.

The pallet arbor is turned and pivoted, and fitted in its holes, the back plate being sawn and cut away for freedom. The pallets can next be roughly filed out to shape by a templet made by cutting out the drawing of the escapement.

The pallet centre hole is opened out to go on the collet on the pallet staff just friction-tight and the pallets tried in the plates, being finally filed up to shape and smoothed to a perfect depth. They are then hardened dead hard, smoothed with emery buffs and polished on at least the acting faces, and finally fixed on the staff. The exact "depth" can be adjusted by drawing the screw holes in the back cock, and when correct the steady pins may be fitted.

The dial plate and feet are either bought or made, and fitted to the frame, the motion wheels mounted, etc., and finally the plates are smoothed by filing, emery cloth, bath-brick dust and water, and finally pumice or globe polish.

Such is briefly the process of making a simple clock. It necessitates much careful work, fine turning and smoothing of the pinions and pivots, but is all within the capacity of a very modest workshop and a small lathe. Many amateurs have made successful clocks that compare with the best purchasable.

Skeleton Clock-making.—The skeleton clock described in the first chapter is made on the same plan as the above, but entails rather more filing work. The castings for the plates must be first hammered flat, then the edges inside and out, all filed true and square. Lastly the flats are filed, the pillar holes drilled, etc., and the train run in. Chap. I. gives many details of construction which in conjunction with those above will enable the clock to be made without difficulty.

Striking Clocks.—The regulation way of making a striking clock is to purchase the castings and forgings as here illustrated, but there is no real necessity to follow the time-honoured custom and make all racks, rack hooks, warning levers, etc., of forged iron or steel. In French clocks these are nearly always of brass and act perfectly, indeed brass has several distinct advantages for these parts; it is not liable to rust, and there is less friction between polished brass and steel than there is between steel and steel. Thus with brass parts, the lifting pins cause less resistance and not so liable to stick. There is also a saving of labour in making, and brass is always procurable, while sheet steel is generally not so easy to get of just the thickness required.

Lifting pins, warning pins, stop pins, etc., should always be of steel and screwed in tight and riveted.   Pins merely driven in generally work loose in a short time.

Rack-making presents a little difficulty to amateurs sometimes.   Machine-cut racks can be bought, but it is quite easy to scribe an arc of a circle for the points of the teeth from the rack pivot hole as a centre (see Fig. 108) and step it with dividers into the required number of equal divisions. The teeth are then filed with a half-round file to an even depth. Of course little inaccuracies occur, but these do not make any difference, as each step of the snail is marked from the rack teeth in the following manner.

The rack teeth are filed and smoothed, the tail riveted on, the rack mounted on its stud, the rack hook made, smoothed

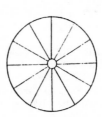

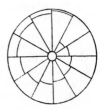

FIG. 108. — Making a Rack.   FIG. 109. — Blank for making a Snail.   FIG. 110. — Marking the Steps of a Snail.

and mounted to fall properly into the teeth.   A blank disc of brass for the snail is mounted upon the hour wheel and divided into twelve portions with compasses and lines scribed as in Fig. 109.

The rack, hook, hour wheel and snail are then put on in position and a sharp-pointed steel pin put in the rack tail. The hook is placed in the first tooth of the rack as for striking "one," and by turning the snail through one division, the pin in the rack tail may be made to scribe an arc of a circle upon it representing the one o'clock step.   The hook is then placed in the next rack tooth and the two o'clock step scribed, and so on. By this method each step of the snail corresponds with its rack tooth, however unequally the teeth may be spaced.   The scribed snail will have the appearance of Fig. 110 and it will only remain to file it to the arcs accurately, afterwards testing each one with its rack tooth as a check.

On a similar plan, a locking plate can be marked out in the clock itself. The locking plate is mounted upon a wheel, having as many teeth as there are strokes in twelve hours. Thus a locking plate to strike hours only must be carried by a wheel of seventy-eight teeth. One to strike half-hours as well, by a wheel of twelve more teeth = 90. The train is arranged so that the pallet wheel makes one revolution for each blow struck.

So to mark out a locking plate, mount the blank in place, put the pin wheel and pallet wheel in the frame as well, and mount the locking arm in position. Lock the pallet wheel by its stop pin as usual at the completion of striking, and mark the locking plate blank where the locking arm comes. Revolve the pallet wheel once and make another mark, revolve it twice and make a third mark, and so on up to twelve. If half-hours are wanted, revolve the pallet wheel once between each hour. It will be observed that the one and two o'clock spaces run together and are cut out as one in an hour-striking clock, while in a half-hour striking clock, half-past twelve, one, and half-past one, all form one long space, being each a single stroke.

Striking Trains.—These are calculated on quite a simple plan. In a rack clock, the pin wheel generally has eight pins to lift the hammer. There are in an hour striking clock 156 blows per day, which equals $19\frac{1}{2}$ turns of the pin wheel, so the main wheel must have a sufficient number of teeth to keep the clock going for the week. In a grandfather clock the pin wheel pinion is eight and the main wheel generally eighty-four or $10\frac{1}{2}$ times the pin wheel pinion. As the line allows of two turns of the barrel a day, there is a margin to spare, allowing the clock to be struck round to set it to time occasionally and ensuring that the striking does not run down before the going, which it should never do.

Next, the pallet wheel must turn once for each pin that passes. Eight is the usual number of pins, and when that is the case the pin wheel must have eight times as many teeth as the pallet-wheel pinion, so that the latter turns eight times for once of the pin wheel. In the grandfather clock, the pin wheel is sixty-four and the pallet pinion eight.

It is immaterial from the point of view of correct striking, how many turns the warning wheel makes to one of the pallet wheel, but it *must* make an even number. That is to say, the pallet wheel must have exactly seven times, eight times, nine times, or ten times as many teeth as the warning wheel pinion, and not $7\frac{1}{2}$ times or any other fraction. This is to ensure that the warning pin will always be in the same position when the train stops. In the grandfather clock, eight times is often adopted and sometimes nine times. When the latter, the pallet wheel is sixty-three and the warning pinion seven. The number of the warning wheel and fly pinion is quite immaterial, but when the pallet wheel of sixty-three is adopted it is usually fifty-six, though for no very particular reason, it might just as well be fifty-five or fifty-seven as far as efficiency is concerned, nothing depends upon it.

One consideration, and one only, governs the actual numbers of the pallet and warning wheels and the warning and fly pinions, and that is the speed that the clock is required to strike. With high numbered wheels, the fly will make more revolutions to one of the pallet wheel than it will if low numbers are adopted. The numbers given above, sixty-three and fifty-six with pinions of seven, give seventy-two revolutions of the fly for each blow struck, and make for fairly slow striking. If fifty-six and forty-two were adopted, the fly would only make forty-eight revolutions, and the clock would strike much more quickly. For a bell clock, fairly quick striking is suitable, but for a deep-toned gong or a tube, slow striking is preferable, and for the latter a ratio of ten to one for the pallet wheel and warning pinion and nine to one for the warning wheel and fly pinion will give good results. This would mean a pallet wheel of seventy and warning wheel of sixty-three, and give ninety revolutions of the fly. If high numbers such as these are adopted, the pallet and warning wheels must have teeth a little finer than the pin wheel so as to keep down the diameter of the pallet wheel to a trifle less than that of the pin wheel.

Going Trains.—In designing a going train, the central fact to keep in mind is that the "centre pinion" must turn once per hour to carry the minute hand upon its arbor. If the clock is to go eight days and have a weight and line, the barrel may make two turns per day as in a grandfather clock, and have sixteen turns of line upon it. Also in an eight-day dial with fusee, the fusee has sixteen turns of chain or line and turns twice per day. Then obviously the main wheel must have twelve times as many teeth as the centre pinion. With a centre pinion of eight, the main wheel is therefore ninety-six.

If the clock is to have a seconds hand upon the escape wheel arbor like a grandfather clock, the escape wheel must turn once per minute, that is it must make sixty revolutions for each one of the centre pinion. Sixty is equal to $7\frac{1}{2}$ times eight, so the usual practice is to use these two ratios, which give a third wheel a little less in diameter than the centre. With third and escape pinions of eight, this gives a centre wheel of sixty-four (8 times 8) and a third wheel of sixty ($7\frac{1}{2}$ times 8), and is the usual grandfather clock train.

Then if the pendulum is a long one beating seconds, the escape wheel must have thirty teeth, as one tooth passes for every two vibrations of the pendulum. If the pendulum were shorter and made eighty vibrations per minute as in some Vienna Regulators, the escape wheel would require forty teeth. If the pendulum made ninety vibrations, the escape wheel would want forty-five teeth, and if the pendulum were 9 in., making 120 vibrations, the escape wheel would want sixty teeth. These two last numbers, however, would be too high for practical use, so the rule is that clocks with such short pendulums are not provided with seconds hands. Then there being no necessity to make the escape wheel revolve once per minute, the high ratio required in the train may be equally divided between all the wheels.

Thus suppose a half-seconds pendulum to be used and a thirty-tooth escape wheel. The escape wheel would revolve twice per minute or 120 per hour. One hundred and twenty is made of ten times twelve, and these two ratios may be adopted, giving a centre wheel of ninety-six and a third of eighty with pinions of eight. Or a centre wheel of eighty-four and third of seventy with pinions of seven.

These examples will show the governing principles of train designing, but any odd train may be calculated as follows :— Multiply together the centre, third and escape wheels, double the result and divide by the third and escape pinion. This gives the number of beats per hour of the pendulum, which again divided by sixty equals the number of vibrations per minute.

As an example take the train : centre, eighty-four ; third, seventy-eight ; escape, thirty-two ; pinions of seven. This is a common train for an eight-day dial.

$$\frac{84 \times 78 \times 32 \times 2}{7 \times 7} = 59904 \text{ beats per hour}$$
$$\text{or } 143 \text{ per minute.}$$

This would require a pendulum of about 7 in. from point of suspension to centre of bob (see method of calculating the length of a pendulum).

In the same manner any train may be worked out.

Suppose the case of an eight-day clock with going barrel. As the number of efficient turns of a mainspring is limited to four or five, an intermediate wheel will have to be interposed between the barrel and the centre pinion, or else the barrel would require an impossible number of teeth. Say four turns of spring are available ; then the barrel will turn once in two days, which is forty-eight times slower than the centre pinion. Forty-eight is made up of six times eight. So one way is to give the barrel eight times as many teeth as the intermediate pinion and the intermediate wheel six times as many teeth as the centre pinion. The numbers may then be : barrel, ninety-six ; intermediate wheel, sixty ; pinion, twelve ; centre pinion, ten.

High-numbered Pinions.—A pinion of six leaves is of very limited use in clockwork. For a minute pinion to drive an hour wheel it may act very well, and for a fly pinion it is passable, but as a transmitter of power for use in a going or striking train, it has great drawbacks. The motion is uneven, the power variable according to the position of the leaf doing the work, and the friction is harsh.

A pinion of seven leaves is the lowest permissible in trains, and even then the friction is rather great and the power variable. Pinions of eight leaves are the lowest solid pinions that run really well. Pinions of ten and twelve are better still. Over twelve leaves are seldom seen except in a regulator occasionally, and over sixteen leaves are of doubtful value. There is not much advantage really in using pinions of more than twelve leaves. Lantern pinions run more smoothly than solid ones, and a lantern of eight leaves is as good as a solid pinion of ten or twelve.

For a going train of a grandfather clock with pinions of twelve, the numbers would be : main wheel, 144 ; centre, ninety-six ; third, ninety.

The Sizes of Wheel Teeth.—It is the practice in English dial clocks and some others to make the entire train with one size of tooth. This is a great mistake, as the greater the pressure, the larger and stronger the teeth should be, and the lighter and more quickly moving wheels ought to be the smallest in diameter. On this plan, the main wheel teeth ought to be coarser than the centre and third wheels. Clock wheels are calculated on " diametral pitch." That is, so many teeth to the inch diameter. This is a convenient system, as the diameter of a wheel with a given number of teeth is easily arrived at. Thus, with teeth of thirty-six diametral pitch, a wheel 1 in. on the pitch circle has thirty-six teeth, $1\frac{1}{2}$ in. has fifty-four teeth, 2 in. has seventy-two teeth, and so on.

Thirty-six pitch is a good one for the main wheel of a weight-driven eight-day striking clock of grandfather pattern. The teeth are large and strong, and for a main wheel of ninety-six gives a diameter of $2\frac{2}{3}$ in. Forty pitch is a good one for the centre and third wheels of the same going train, or for the pin wheel of the striking train. Forty-two pitch is suitable for the pallet and warning wheels.

Forty-two is the pitch found in most English eight-day dials, though they would be better with main wheels of forty and the rest forty-four, on the principle that the largest teeth are required where there is the greatest pressure.

Chiming Clocks.—The train of the chime mechanism of a quarter-chiming clock is designed to revolve the pin barrel twice per hour, and the pallet wheel ten times per hour, that is, once for each chime. This is usually done by making the second wheel of the chiming train of eighty teeth, driving a pallet pinion of eight. Then the second wheel revolves once for ten times of the pallet wheel, or once per hour. This is the same speed as the centre wheel, and to go eight days it is obvious that a main wheel of ninety-six and pinion of eight will serve as in the going train, giving a ratio of twelve to one, with sixteen turns on the chime barrel. But the chiming may be repeated in putting the clock to time occasionally, and should never be allowed to run down before the going, therefore it is usual to give the chime main wheel one hundred teeth, being a safety margin of four.

The numbers of the pallet and warning wheels may be the same as in a striking train, and for the same reasons. It is seen from the above that the second wheel of eighty turns once per hour. This is made to drive the pin barrel by means of a wheel of forty mounted on the pin-barrel axis. The pin barrel thus turns twice for each revolution of the second wheel and makes $\frac{1}{5}$ of a revolution for each turn of the pallet wheel. As each fifth part of the pin barrel contains one chime, this effects the purpose desired. In tube-chime clocks the second wheel is often one hundred and a wheel of fifty on the pin barrel, with a pallet pinion of ten.

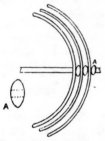

FIG. 111.—Mounting Chime-bells.

Making a quarter-chiming clock presents no great difficulties, being merely a question of time, labour, and patience. Many excellent chiming clocks have been made by amateurs, and go perfectly. The racks and snails are marked out as before described. The bells (see Fig. 94) are mounted upon a steel spindle and separated from each other by small wooden washers shaped as shown in Fig. 111, A showing an enlarged washer. They are rounded in this form so as to touch the bells only in the centre. If flat they would seriously strain the bells when tightened up and also partially stop the tone by holding them too rigidly.

The hammer rack also shown in Fig. 94 is a difficulty. It consists of a bar of solid brass about $\frac{5}{8}$ in. wide and $\frac{1}{4}$ in. deep, screwed across the clock frame and slotted for the hammer joints to work in. A pin goes through it from end to end and serves as the joint pin for all the hammers. To drill such a long hole of so small a diameter would be almost an impossibility, so it is accomplished thus. A saw cut is made from end to end of the rack, as shown in the end view

(Fig. 112). C is the brass block, the saw cut is shown at A. A slip of brass is filed up to fit the cut and soft-soldered in, leaving the hole D for the joint pin to pass through. B is a hammer tail.

When the crown-wheel system of driving is used for a tube clock, as shown in Fig. 95, the crown wheel is squared on to the projecting arbor of the eighty wheels of the train. This arbor should be left stout and strong so as not to spring unduly. For this reason, in such clocks it is advisable to give the second wheel a fairly large pinion, as then the solid core will be stouter. This means a large main wheel and is yet another argument for large main wheel teeth in these clocks.

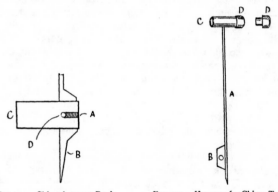

FIG. 112.—Chime-hammer Rack.   FIG. 113.—Hammer for Chime Tube.

Thirty pitch will not be too large for this wheel and pinion. The crown wheel may have eighty teeth, but one hundred is better. If eighty, the wheel on the pin barrel must be forty. If one hundred, the pin barrel wheel must be fifty. This wheel of fifty should be placed in a chuck and have its entire centre cut out, leaving only a ring with teeth. The toothed ring can then be slipped tightly over the tube of the pin barrel and soft-soldered on. It can come between two hammers and not be in the way.

Hammers for tubes should be made as in Fig. 113. The portion working in the hammer rack may be of brass, and the steel wire may be soft-soldered in, a hole being drilled through.

L

Or the flat portions may be steel with the wires silver-soldered or brazed on the edge. The head is a piece of brass tube about 1 in. long and ½ in. diameter, screwed on the end of the hammer stem. B is the flat portion working in the rack, A is the steel wire stem, C is the brass tube, and D is a turned wood plug fitting the tube and forming the face of the hammer. Stretched tightly over the hammer face and tied there should be a piece of rubber cloth to soften the tone. The hammer stems are usually about 6 in. long and supplied with *very* light thin brass springs.

Chime tubes should be hung with very short cords, the tube tops being as close as possible to the supporting bracket without touching, as in Fig. 97. Then the tubes do not swing much when struck. The hammers should strike about half-an-inch from the top.

Hammers for circular steel gongs are generally made of steel wire, so that the heads can slide along them and be clamped by a small set-screw at the point where the best tone is obtained. A hammer head for a large gong (6 or 7 in. diameter) may be a block of brass ¾ in. square and ¼ in. thick, as in Fig. 114. C is the hammer stem, B the clamping screw holding the head in place, and A is a piece of hard leather screwed into a hole in the hammer face. The leather produces a nice soft tone.

Date Work.—Simple date work as found in grandfather clocks and bracket clocks consists of a disc of metal behind

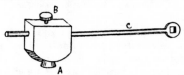

the dial, having the numbers 1 to 31 painted around it. In the dial is a small opening through which one number may be seen. It is necessary for the dial to turn one

Fig. 114.—Hammer for Circular Wire Gong.

division each twenty-four hours just at midnight. It is accomplished thus. Upon the hour-wheel another wheel is mounted, which drives an intermediate wheel turning on a fixed stud in the front plate. The disc has thirty-one teeth and a "jumper" like that on the star wheel in Fig. 83. A pin fixed upright in the intermediate wheel engages with the teeth of the disc and turns it round one tooth. The wheel mounted on the hour wheel has forty teeth, and the intermediate wheel eighty. Or they may have thirty and sixty. The effect is that the intermediate wheel turns once in twenty-four hours and so moves the date-disc at midnight. This date work will of course only be correct for the thirty-one-day months. At the end of the other months it will require setting.

Perpetual calendar work, though seen in a fair number of watches, is in clocks quite a curiosity. It is somewhat complicated and highly ingenious. By its means, the date is shown correctly month by month and year by year, leap year included. It depends for its action upon a " month wheel," a wheel that revolves once in four years, and carries a sort of " snail," the steps of which are cut to varying depths to correspond with the long and short months. Upon the steps a kind of rack falls, and the depth regulates the number of divisions turned by the date wheel before commencing again at " one."

Moon Work.—The usual moon disc has two moons painted upon it and turns once in two. lunations. A lunation is approximately 29½ days. So if a moon disc with two moons has 118 teeth and is moved one tooth each twelve hours by a pin in a wheel gearing with the hour-wheel and of the same size, it will only have an error of about three-quarters of an hour each month, which in a year amounts to nine hours and in three years to about a day, which is quite near enough for most clock owners. A train that is nearly exact is as follows (Lord Grimthorpe) :—A pinion of twelve on the centre pinion (turning once per hour), drives a wheel of 182 with a pinion of nine on its axis. This pinion drives another wheel of ninety-one with a pinion of thirty-seven on its axis, this last drives a wheel of 171, which will turn in 29 days, 12 hours, 44 minutes, 3·4 seconds, and has an error of half-a-second per month.

Equation of Time.—The equation of time is the difference between twelve o'clock local noon (shown when the sun is due south at real midday) and twelve o'clock noon meantime, shown by the clock. It is the amount that the clock is " fast or slow of the sun." At some times of the year both coincide, at others there is a difference of nearly a quarter of an hour. In the few clocks showing it, a wheel is arranged to revolve once per year, and may have 365 teeth

moved one tooth per day like a date-wheel. Upon the axis of
this wheel is a cam so shaped that a lever resting on its edge
rises or falls as the wheel goes round. The cam is cut so that
a hand attached to the lever shows zero at those dates when
the sun and clock agree, and falls below zero when the clock
is behind the sun, and rises above it when the sun is behind
the clock. A scale divided into minutes and fractions in
conjunction with the hand, shows the amount of the difference
in time.

Quite obviously this arrangement also offers a solution of
the " perpetual calendar" difficulty. A hand attached to the
axis of the " year-wheel" will revolve once per year, and a
circle at the dial edge may be divided into months and days,
the hand showing the correct date except in February, leap
year. The disadvantage is that the day divisions are small
($\frac{1}{365}$ of the circle), and require closely looking' at. Still a fair
number of clocks were so made, though possibly none are
made to-day.

Sidereal Time Clocks.—Sidereal time is required in
observatories. The sidereal day is $\frac{1}{366}$ of a year and is
3 mins. 56 secs. shorter than a solar day. To show sidereal
time upon an ordinary clock as well as mean time can only be
done approximately. This coupled with the fact that those
who require sidereal time do so for making astronomical
observations and must have it exact to a second, renders such
clocks really useless. Consequently sidereal time is kept in
observatories by separate regulators, whose pendulums are
so regulated as to beat sidereal seconds, their construction in
all respects resembling mean-time regulators.

Regulators.—To keep accurate mean time for regulating
purposes either in watchmakers' shops or in observatories,
regulator clocks are used. A regulator is a timepiece with a
compensated seconds pendulum, all made as simply and as
accurately as possible. Everything is subordinated to timekeeping.

An average regulator would be made with well-polished
pinions of twelve, fourteen, or sixteen leaves, with fine pivots,
sometimes working in jewel holes. The pallets would be
dead-beat of the pattern shown in Fig. 63, and jewelled where

the escape teeth touch them like Fig. 66.   It would have
maintaining work to keep it going during winding.   The
frame should be heavy and the pillars stout, to prevent
vibration or springing.

To ensure solidity the pendulum is generally hung upon a
strong cast bracket screwed to the case back, of the pattern
shown in Fig. 115.   The suspension is made after the fashion
of Fig. 116.   A steel spindle A passes through the top of the
suspension spring and is screwed up tight by the nut B.   The

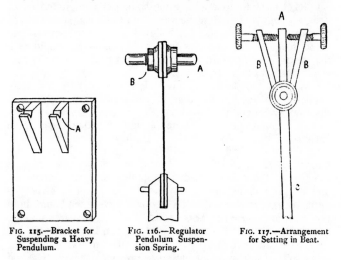

FIG. 115.—Bracket for
Suspending a Heavy
Pendulum.

FIG. 116.—Regulator
Pendulum Suspen-
sion Spring.

FIG. 117.—Arrangement
for Setting in Beat.

spindle A rests in the V slots on the bracket (Fig. 115).   This
should be about 6 in. × 4 in. and $\frac{1}{4}$ in. thick.

To set a regulator exactly in beat, the crutch is generally
made movable upon the pallet arbor and held between two
beat screws, as in Fig. 117.   A is an upward extension of the
crutch C.   BB are two arms rigidly fixed to the pallet arbor.
Two thumbscrews working in BB hold A centrally, and
turning them a little will move A slightly and alter the beat.
In making this arrangement, care should be taken to make the
crutch and beat screw arms, etc., extremely light so as to throw
as little weight as possible upon the pallet pivots.   As a rule
crutches of regulators are far too heavy.

Sometimes the gravity escapement is used in a regulator, but it is doubtful if it has any advantage. The heavy driving weight required is a disadvantage in that it causes much more wear and tear.

The hands of a regulator should be extremely light. They are best filed out from thin French clock mainspring, and the seconds hand, from watch mainspring.

Some regulators in observatories have pendulums as heavy as 40 lbs. This is all very well when they are suspended from a bracket bolted into a solid stone wall, but if mounted in an ordinary regulator and suspended from the back (1 in. wood) of a case that is held to a brick wall by two wood-screws into wood plugs in the ordinary way, such pendulums would rock the clock, and better timekeeping is got from more moderate weights, say from 10 to 15 lbs.

An invar pendulum of this weight may have a rod $\frac{1}{4}$ in. diameter, threaded at the bottom with a thread of eighteen to the inch (Whitworth $\frac{5}{16}$ thread). Then one turn is approximately 1 minute per day. The rating nut should be a large flat disc of brass 2 in. diameter. Upon it and turning with it, may rest the cylindrical cast-iron bob, 8 in. high and $2\frac{3}{4}$ in. diameter. This makes a 13 lb. pendulum. A brass cap on the top of the bob is divided on its edge into sixty divisions, and a clip pointer sprung on the rod just over it serves to indicate any movement. To regulate such a pendulum, hold the rod and turn the bob round, each $\frac{1}{60}$ division making 1 second per day difference. Fractions can be estimated.

The cast-iron bob may be turned smooth and bright and nickel-plated or simply lacquered to prevent rusting. The rod should be polished and just left greasy from an oiled rag. The error of such a pendulum, when first-grade invar is used, is hardly perceptible.

A good plan is to make the bob with two brass caps each $\frac{1}{4}$ in. thick. The bottom cap is tapped and serves as the rating nut. The top cap is bevelled on its edge, French silvered, and divided to form the scale with pointer. Each cap is fastened to the bob by two screws. The bob itself has a $\frac{1}{2}$-in. central hole and the top cap is bored to accurately fit the rod, so that the cap forms the bearing upon the rod, the cast iron not touching it. This simplifies the construction and ensures smooth and easy turning. Fig. 70 shows a pendulum made thus.

Making Pallets.—In drawing an escapement the escape wheel should be laid upon the sheet zinc, or white card and traced with a fine point. One tooth should lie on the vertical centre line. Then if of thirty teeth, mark four spaces on either side and draw tangents so as to cut the centre line above the escape wheel and mark the position of the pallet centre, as in Figs. 23 and 63. If the escape wheel has fewer teeth, the pallets will embrace less than eight. If high numbered, they will embrace more, but in any case, tangents to the points where the pallets engage will mark the position of the pallet holes.

This done, with recoil pallets like Fig. 23, a vertical and a horizontal line (A and B, Fig. 23) will indicate the form of the pallet faces.

In a dead-beat escapement like Figs. 62 or 63, the exact angle of impulse is important. It should be $1\frac{1}{2}°$ on each pallet. The angle of impulse is measured from the pallet hole by means of a protractor. It is the angle included between the dotted lines C and D (Fig. 62) and determines the slope of the impulse faces and angular distance through which the escape wheel drives the pallets.

The outline of the pallets should be very carefully made clear upon the drawing of the escapement. The depth or position of the pallet hole can be transferred first to the clock plates and the pivot holes drilled, the pallet arbor filed and made ready to receive the pallets. Then the pallet pattern may be cut out and used as a template from which to file up the pallets. As they approach nearly to shape and dimensions, they may be tried in the clock and proceeded with very cautiously until finished. After hardening, a little more shaping and trimming can be done to finally fit them, by means of emery buffs or by a soft steel polisher and oilstone dust as in pivot smoothing, and the locking faces and impulse faces finally polished in the same way with red-stuff and oil.

It should be remembered that the locking faces of dead-beat pallets are true arcs of a circle struck from the pallet pivots as a centre.

CHAPTER XIII.

ALTERING CLOCKS.

Silk Suspension Pendulums.—Many old French eight-day clocks have the pendulums suspended by a loop of silk thread. This provides a ready means of regulation, one end of the thread being merely wound up on a steel arbor by a thumbscrew, thus shortening the loop. The pendulums of these clocks are generally very light, and the rods simply thin steel wires ending at the top in a small hook to hang upon the silk loop.

One might suppose that this form of suspension would be almost frictionless, but in reality there is more friction than with the ordinary suspension spring, and the silk thread has the further disadvantage of being affected by the weather and causing inequalities in the rate of the clock. For these reasons many silk suspension clocks are converted to the ordinary form. The best way to convert one is to purchase a " Brocot " suspension and fit it to the back cock, bringing the regulating arbor through the dial in front, just above the 12 o'clock. This also means a new heavy pendulum and altering the crutch.

Another way is to screw a stout brass stud into the back cock, slot it to take the top end of the suspension spring, and fit a new pendulum having the regulating screw at the bottom. The disadvantage of this plan is that no regulating can be done from the front.

Old Verge Bracket Clocks.—There are still a few old English bracket clocks about with verge escapements, and in all of them that have been kept continually running the escapements are completely worn out. The verge escapement, with its light pendulum, knife-edge suspension and large arc of oscillation, is incapable of keeping good time, and unless the clock is to be kept intact as a specimen of some particular old maker's work, it is advisable to convert such clocks to the modern form. This entails a good deal of work, and includes mounting and fitting a new third wheel, escape wheel and pinion, pallets, crutch and pendulum, as well as filling up many old holes in the plates and making a back cock.

The new third wheel may have the same number of teeth as the old crown wheel, and if the new escape pinion has the same number of leaves as before, the new recoil escape wheel may have two more teeth than the verge wheel, the object of which is to use a rather shorter pendulum, or else the rating nut will touch the bottom of the case.

The escape wheel and pinion will be planted upon the centre line of the frame, and the third wheel and pinion will have to be moved to the right considerably, as in an ordinary English clock, so as to depth with the centre wheel and escape pinion.

A casting for a back cock can be obtained for a few pence, and a complete pendulum may be purchased ready-made if desired. The pallets will be filed up from an ordinary pallet forging.

If it is desired to make a pendulum instead of buying one, the rod may be made from a strip of steel of the usual flat section, or brass may be used, but has a larger temperature error. A bob of the ordinary brass shell pattern filled with lead presents difficulties, but a good bob may be simply made from three discs of brass, one the thickness of the rod and the other two rather thicker. Discs stamped out from sheet brass, of any ordinary diameter such as 2 in., 2¼ in. and so on, can be purchased at the clock founder's. Three of these, making in all about $\frac{3}{8}$ in. in thickness, will make a good bob. Divide the centre one with a saw and remove a portion equal in width to the rod. Then rivet them all up together with four brass rivets and smooth off on both sides and the edges. This makes a solid brass bob that will slide upon the flat rod perfectly.

Substituting Chains for Lines in Fusee Clocks.— Fusee clocks with gut lines are continually giving trouble.

The lines break and fray, and whenever a new one has to be fitted, it is necessary to take the clock to pieces. A chain very seldom breaks and it is a simple matter to make the alteration. Cutting the hollow in the fusee to accommodate the chain hook is the principal difficulty, and this may be overcome with a little patience. A flat is first filed out, then several holes are drilled close together to deepen it, and the intervening brass cut out by a small chisel made from a broken graver. The hollow should be crescent-shaped and deep enough for the hook to lie in flush. A steel pin should be put across it for the hook to hold on to.

The hole in the barrel for the other chain hook must be well undercut so that the hook may get a good hold and lie flat upon the barrel surface. In some old clocks, the fusee groove is round-bottomed and shallow. Such a fusee will need the groove re-turning before a chain can be fitted, but most clocks will be found to have a square-bottomed groove whether they have lines or chains, and this is perhaps one of the causes of lines fraying so readily, a round-bottomed groove not chafing the line so much.

Making a Grandfather Clock Strike on a Gong or Tube.

—A circular steel gong is seldom very satisfactory when fitted to a Grandfather clock. As a rule there is not sufficient room for it ; its proper fixing presents difficulties, and neither the case back nor the seatboard makes a good sounding-board.

On the other hand, a long tube, like one of those in a tube chime clock, may be fitted and give perfect satisfaction. It should be suspended from a support about 2 in. above the seatboard and hang through a hole cut to allow it free passage. A new hammer must be made to hang downwards and strike outwards as in Fig. 95. An addition to the driving weight will be necessary and the amount will depend upon the weight of the hammer. The hammer being made and the tube hung, additional weights may be added until the striking is prompt. The whole may then be weighed and a new weight cast of the required dimensions. The writer has known a 4-ft. length of 1-in. steel bicycle tube used with good effect.

Adding Quarter-Chimes to a Grandfather Clock.

—Cases of these clocks differ very much in their dimensions.

Those with narrow waists will not accommodate the three weights necessary, especially when it is remembered that the chiming weight must be of large diameter. The length of the chiming weight requires a case of more than average height, that it may have sufficient fall to go for eight days. Given a case high enough and wide enough, quarter-chimes on four and eight bells may be added. Tube chimes cannot possibly be got into an ordinary case, as the usual inside depth from back to front is 6 in. and a tube clock requires at least 9 in.

The conversion is best carried out by making a complete new clock frame with plates about 8 in. square. The old going train is pitched in the centre, the striking train to the left, and the new chiming train to the right. The cannon pinion, motion work, hour wheel, and snail are all used as before, and the striking rack. A quarter snail is made and mounted upon the minute wheel, a new hour rack hook, and warning lever after the pattern shown in Fig. 93, and quarter rack, hook, and warning lever, etc., fitted. Fig. 94 shows the hammer work.

It is of course possible to mount a chiming train in a separate frame and bolt it on to the side of the old movement, but it makes a poor job and entails nearly as much labour.

Improving the Timekeeping of a Grandfather Clock.—A Grandfather clock can be made into a sort of semi-regulator and greatly improved as a timekeeper by converting the escapement to a "dead-beat" and fitting a new pendulum with wood rod and lead bob. The backs of the cases are generally the weakest, and when this alteration is decided upon, it will be wise to strengthen the back with an additional $\frac{3}{4}$ in. of sound deal board secured by a dozen or more screws to the old back. The clock may then be securely fixed to a wall and the pendulum may be hung from a bracket on the case back after the style of Fig. 115. The wood rod may be pine $\frac{1}{2}$ in. diameter and of round section. Brass tube forms a good end for top and bottom. A cap 1 in. long may be fitted tight and flush, and pinned through at each end. The top cap has the suspension spring attached and the bottom cap the rating screw. The bob may be oval or cylindrical, about 7 in. high and 2 in. diameter. The crutch should be widened to span the wood rod, by having a new fork brazed on to its end. Where it works on the rod, a short length of brass tube should be slipped over like a ferrule.

In altering the escapement, either the pallets must be raised about $\frac{1}{8}$ in. or the new straight-tooth escape wheel must be rather smaller than the old one. The pallets will be of the form shown in Fig. 63 and can be filed up from a forging, or if desired may be made like Fig. 62.

The error of a clock so altered will not exceed five or six seconds per week.

Thirty-hour Grandfather Clocks.—The question is often asked, " Can these clocks be altered so as to go eight days ? " They cannot, except by fitting a fresh movement to the old dial, or by completely remaking the old one. It would require two new modern barrels, new centre wheel and pinion, new third wheel and pinion, new motion work and a rearrangement of the striking train. It would probably be found best to fit a new or second-hand eight-day movement complete.

CHAPTER XIV.

ELECTRIC CLOCKS.

UNDER this head come all those clocks in which electricity plays a part. In some electricity is the actual motive power, in others it is merely employed to rewind the clock at regular intervals. Or again it may be employed to "control" a clock either by influencing the pendulum or by setting the hands at each hour. A system of clocks may be driven from one "master clock" or regulated by electric impulses each half minute or more frequently. In this case, the driven clocks merely consist of a few wheels and a pair of hands.

Electricity can also be applied to ordinary clocks for the purpose of operating alarms, chimes, etc., or for striking a bell each second or minute for counting purposes. A brief description will be given of these different kinds of electric clocks, though as far as the electric part is concerned it is electricians' work, and it is not within the scope of this book to give directions either for making or repairing them. Also it should be noted that nearly all the useful and successful electric clocks proper are covered by patents at present, as the whole subject is yet in its infancy. There are great possibilities and endless opportunities for the clockmaker-electrician to exercise his ingenuity. It is quite certain that clocks on these lines have a large future before them.

Electrically-driven Clocks.—A few early electric clocks consisted of a train of wheels driven by a pendulum, which at each beat caused the seconds wheel to advance one tooth. The train consisted of just sufficient wheels to work back from the seconds wheel to the minute and hour wheels and so move the hands. They had no barrel, mainsprings or weight, the wheels simply running lightly and freely with no power upon them except the very slight forward impulse given to the seconds wheel at each vibration of the pendulum.

The vibrations of the pendulum were maintained by magnetic attraction, an electro-magnet being arranged beneath the "bob," the current being obtained from the earth by burying the wires deeply in such a way as to ensure the constant presence of moisture. The current so obtained is weak, but sufficed.

The system had many weak points and not many clocks were so constructed. The rate of such a clock would vary according to the strength of the current, the pendulum being almost constantly controlled by the magnet.

The central idea of this clock was that of driving the pendulum direct, and operating the hands from the pendulum, and at the same time producing a clock that would never want winding, but would run uninterruptedly as long as the train kept in condition and the electric current was available.

A modern way of accomplishing this is to drive the pendulum by a kind of gravity escapement, in which an impulse is given at regular intervals by the falling of a weighted arm through a given space. In these clocks, the electric current is merely used to raise the weighted arm in position again ready for the next impulse. This frees the pendulum from all errors due to variation in the electric current and gives it a constant impulse and a constant arc of vibration. It also leaves it quite free (as far as the driving power is concerned) except just at the moment of receiving an impulse.

A pendulum driven thus and operating a train of wheels connected with a set of hands is termed a "Motor-Pendulum." It is given a certain amount of work to perform, viz. to drive the train. At each return vibration, an arm upon its rod has to turn a ratchet-wheel one tooth, and the ratchet has to be held in position by a "jumper" of some kind answering the same purpose as that on the star wheel of a striking clock. This means a serious interference with the free vibration of the pendulum, the work it has to do being variable according to the effect of temperature upon the springs of the various parts, or the fluidity of the oil upon their faces and pivots.

In Fig. 118 J is a pendulum rod. E is a dead-beat impulse pallet rigidly fixed to the pendulum rod. C is the

gravity arm working on a pivot at M.   F is a small roller fixed
upon C.   D is a light spring catch, the flat top end of which
holds up the gravity arm C so that the roller F is just clear of
the upper dead face of the pallet E.   When the catch is
released, C falls until the roller F rests upon the dead face of
E.   On the return swing of the pendulum F passes down the
inclined impulse face of E and gives the pendulum a strong
impulse.   On falling below E the gravity arm C is arrested by
the contact screws at B touching.   Instantly the electric circuit
is closed and the electro-magnets L draw A forward sharply,
thus raising the gravity arm C until D catches and holds it up
again ready for the next impulse.

G is a light arm pivoted upon the pendulum rod.   A projecting piece I engages with the teeth of the fifteen-toothed
ratchet H.   For fourteen teeth of the ratchet, each swing of the
pendulum turns the ratchet one tooth, and the end of G does
not touch D, working in the looped portion, and the gravity
arm C therefore remains held up.   The fifteenth tooth of H is
cut deeper than the rest, and I pressing into the space, rises a
little.   Therefore when the pendulum swings to the right the
end of G comes in contact with D and forcing it outwards, C
is released and F falls on the pallet E.   Thus at each revolution of H, one impulse is given to the pendulum.   H having
fifteen teeth, revolves once in 30 seconds, causing the pendulum
to receive two impulses per minute.

Each time an impulse is given the electric circuit is closed.
This gives two alternative ways of working the hands of the
clock.   First H may be made to drive the train mechanically,
or the electric current may be made to do the work by jumping
the minute hand forward each half minute, a very simple little
electrical device only being required for this purpose.

Obviously the same current may be used to drive the hands
of any number of dials in the circuit, and the one motor
pendulum may be used as the master-clock of a system of
synchronized clocks.

It will be noticed that a failure of the current occasionally
for short periods will not cause the motor pendulum to stop, as
its momentum is sufficient to keep it vibrating for some time,
nor will the failure cause H to cease revolving.   So if H works
the train and hands of its own dial mechanically, a failure of
current now and then would not affect the clock noticeably.

On the other hand, if the hands are moved by the current, or if other dials are on the same circuit, all will miss an impulse, *i.e.* a half minute, each time failure occurs.

Fig. 118 is practically the " Synchronome " system.

The " Eureka " electric clock is a small and compact mantel

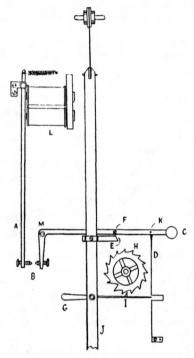

FIG. 118.—Electric Motor-pendulum.

timepiece. It is operated by a " motor-balance," that is, a large balance made on the plan of a watch or chronometer compensation balance. The balance receives a magnetic impulse at each vibration, sufficient to maintain its vibrations and enable it to drive the train, which consists of a very few motion wheels only and takes very little power. So little is required that the batteries can be placed in the base of the clock making it complete in itself.

Clocks Wound by Electricity.—Under this heading come clocks in which a spring or weight is the actual motive power, driving the pendulum through a train and escapement after the manner of ordinary clocks, but in which electricity is used to rewind the spring or weight at regular intervals. Continuous running is thus attained as long as the mechanism remains in running order and the current is sufficiently powerful to do the rewinding.

It is quite obvious to a mechanic or an electrician that an electro-motor can be geared to the winding arbor of any clock and supplied with current to wind up the weight or spring. It is equally obvious that only a little ingenuity is required to make this winding automatic. For instance, the weight having fallen to a certain point may be made to switch on the current, and when wound up sufficiently, to switch it off again, the only necessity being a constant current such as can be obtained from a town supply.

This system would be expensive to make and is probably not found in many clocks. It has been found simpler and more economical to apply the rewinding to a spring much nearer to the escapement and at frequent intervals, rather than to wind the main arbor where only an electro-motor of some power can do the work. The third wheel of a clock, for instance, may be made to revolve in eight minutes, and if it were provided with a large fine ratchet and a driving lever sufficiently heavy to drive the escape wheel and pendulum, a very small amount of power would be required to raise the lever every half minute. In half a minute the wheel would revolve $\frac{1}{16}$ of a turn, and an electro-magnet could be arranged to lift the driving lever through this angle, thus dispensing with an electro-motor and powerful current.

In Fig. 119, A is the third wheel. B is the driving ratchet turning easily on the axis of A and connected with it by a spiral spring of watchspring form, through which the driving is done, the spring acting really as a maintaining spring. C is a maintaining detent to hold B so that the spring may continue

M

to drive while the weighted lever D is being raised by the electro-magnet G. The lever D engages with the ratchet teeth through the click E and drives the third wheel. As soon as D falls to a certain point, the contact screws F complete the electric circuit and H is drawn sharply upwards raising D, which takes up the driving again. And so the clock is driven continuously; each time D falls, the circuit is completed and it is raised again, thus keeping the spiral spring under tension and driving A with a constant force.

A failure of current in such a clock for more than a few minutes means stoppage, but this risk is lessened by giving the

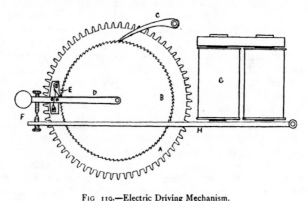

FIG 119.—Electric Driving Mechanism.

intermediate driving spring several effective turns, and also by a system of two driving levers which are raised alternately.

Clocks driven by Electric Impulses.—An electric master-clock actuated by a motor pendulum, like that in Fig. 118, or an ordinary regulator designed to close an electric circuit every half minute, can be made to drive any number of dials by quite a simple electric contrivance, by which a pawl draws a ratchet-wheel round one tooth at each impulse. A clock so made need only have a ratchet of one hundred and twenty teeth upon its centre arbor and the usual motion wheels to drive the hour hand. Such clocks are really only electrically operated dials, and if the contact occasionally fails, the error (half a minute) may be corrected at the termination of each hour by one of the synchronizing systems worked from the master-clock.

A more reliable system is to make the impulses from the master-clock drive the pendulums of the clocks in the circuit by electro-magnetic influence (Ritchie's). The pendulums drive the trains, and should the current fail for a short time, the momentum of the pendulum is sufficient to cause it to continue to vibrate until current is again supplied, whereas a failure, in the case of mere electric dials, if lasting for more than one beat, puts them even beyond the control of the hourly correction. And in electrical work a faulty contact or a failure of current is always a possibility.

Electrically-controlled Clocks.—There is a system by which current received each second, or at other intervals, from a good regulator can be made to influence the pendulum of an inferior clock and keep it in unison with the master pendulum. This is effected by electro-magnets influencing the pendulum itself without any mechanical contact.

But a more usual way of making a master regulator control the time of a number of other clocks, is by an hourly or half-hourly current being applied to simple mechanism, which actually re-sets the hands each hour or half-hour.

Fig. 120 shows a method that was installed some years ago in numbers of clocks at railway stations and in other places. It is made under the assumption that the clock will not have an error of its own of more than half-a-minute either fast or slow, in the hour. Just at the hour, therefore, the point of the minute hand will be over the crescent slot shown in the dial. Behind the dial is a kind of scissors arrangement, having two upright pins AA standing up normally at the top ends of the crescent as shown. At the hour, the current causes the scissors to close, the two pins AA coming towards each other and meeting upon the hand in the centre. Obviously if the hand is not quite in the centre, the pins will set it centrally and so correct the time. This arrangement did not always act perfectly. If one of the clocks stopped of its own accord, it remained stopped, or if it varied from some cause, such as loose hands, etc., more than the prescribed half-minute, it remained wrong. Even if the minute hand come within the radius of the pins AA, it was not always set correctly, as dirt or oil upon them caused stickyness, and sometimes the hand point would adhere to one pin and be drawn back by it, thus actually setting the clock wrong.

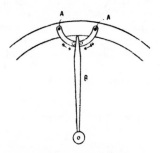

FIG. 120.—Electric Setting Mechanism.

A similar object was attained by another means. The clocks to be controlled were all regulated to gain perceptibly, say a quarter minute each hour. A pivoted lever was arranged to stop the minute hand exactly at sixty mins. and hold it there until the electric current raised the lever and freed the hand, and so on at each hour. This method had advantages over the first (Lund's), but neither are used to any large extent.

Electric Alarms.—An ordinary drum alarm clock may be made to ring an electric bell in many ways. One very simple way is to place two pieces of thin brass 3 in. high, upright in a wood base, and connect the electric wires one to each. The brass strips should be half-an-inch apart. The clock is placed upon the wood base so that its alarm winding handle is between the brass uprights and in a vertical position, not touching either of them, or only touching one. When the alarm goes off, the alarm winding handle runs back and makes intermittent contacts between the brass uprights, ringing the electric bell or bells upon the circuit.

A clock that has not already an alarm is a little more difficult to deal with, as the letting of mechanism has to be made and some means devised to enable it to be set to any required time.

In most clocks there is room for a wood or vulcanite disc to be mounted upon the hour wheel in such a manner that it can be turned round to set to any time. A thin wire across its edge is placed so that it connects by friction with the clock movement. A thin spring resting on the edge of the wood disc will make contact with the wire and cause the alarm to ring. The hour wheel moving slowly, the bell will ring a long time and must be switched off as soon as the alarm has done its work.

Another method often adopted is to prolong the hour hand so that it reaches to the edge of the dial. Then the hour hand must be insulated from the clock by mounting on a vulcanite collet. A friction spring attached to one terminal makes permanent contact with the socket of the hour hand. To complete the circuit it is necessary to connect the hour and minute hands. This is done by arranging two movable metal contact pieces around the dial edge, connected by a flexible wire. At each revolution the minute hand point sweeps over one, but the circuit is not closed until the hour hand is also upon its contact block, and that occurs only at the fixed time once in twelve hours.

All these systems require the current to be switched off during the day-time unless the alarm is to ring morning and evening. By adding another wheel to the motion work a 24-hour alarm could be arranged that would only ring in the morning.

Electric Chimes.—The ordinary grandfather clock has no room for chiming mechanism except by almost re-making, as described in another part of this book, but electricity gives us a means of operating a chiming mechanism from contacts made by the clock, which chimes may be made to sound in any desired part of the house.

The mechanism may take the form of a separate clock-driven chime movement released each quarter of an hour by electric current from the clock, or it may be wholly electrical, each hammer being raised by an electro-magnet actuated from a pin barrel in the clock itself. Both systems give satisfactory results and offer an extensive field for the ingenuity of clockmaker-electricians.

INDEX

PRINTED BY WILLIAM CLOWES AND SONS, LIMITED, LONDON AND BECCLES.

CROSBY LOCKWOOD & SON'S

LIST OF WORKS

ON

TRADES AND MANUFACTURES, THE INDUSTRIAL ARTS, CHEMICAL MANUFACTURES, COUNTING HOUSE WORK, ETC.

A Complete Catalogue of NEW and STANDARD BOOKS relating to CIVIL, MECHANICAL, MARINE and ELECTRICAL ENGINEERING; MINING, METALLURGY, and COLLIERY WORKING; ARCHITECTURE and BUILDING; AGRICULTURE and ESTATE MANAGEMENT, etc. Post Free on Application.

7, STATIONERS' HALL COURT, LONDON, E.C.

AND

5, BROADWAY, WESTMINSTER, S.W.

1913.

Printed in the United States
71192LV00002B/100